GUIDE TO

VIET CONG AMMUNITION

ISBN: 0-87947-111-5

NORMOUNT TECHNICAL PUBLICATIONS
Box 211
Forest Grove, Oregon, 97116, USA

INTRODUCTION:

Having long since passed the stage of logistical development where they are supplied only by what they can afford to manufacture or steal, the Viet Cong currently have a somewhat standardized issue of weapons and ammunition. Although there are still thousands of miscellaneous and improvised weapons in VC use, the present issue is mostly standard Communist Bloc ordnance, and uses the standard Communist Bloc ammunition. This, plus a good assortment of French weapons and some items of Standard NVA or VC manufacture, comprise the usual VC arsenal.

Thus, this volume identifies and describes the standard items of issue for the widely distributed Communist Bloc weapons, and gives complete technical data on such ammunition. With such a myriad of miscellaneous leftovers from numerous wars available to the VC for nonstandard use, it would be impossible to cover the ammunition for all these within the scope of this volume. Such weapons and their ammunition, however, are becoming increasingly scarce in VC use, due to more intensive logistical support from the other Communist Bloc countries.

CONTENTS :

CONTENTS—Continued

CONTENTS—Continued

CONTENTS—Continued

GUIDE TO

VIET CONG AMMUNITION

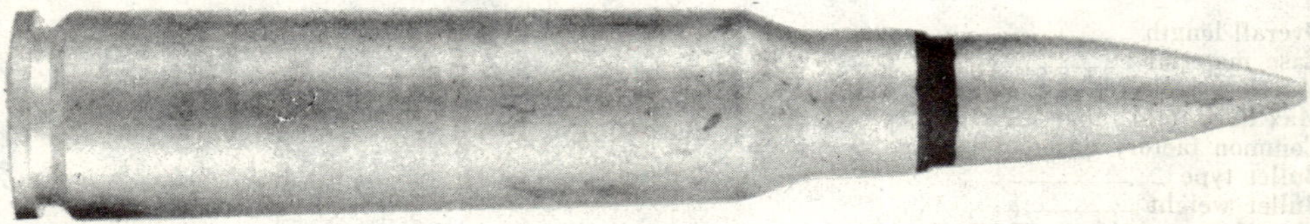

FRENCH 7.5-mm BALL CARTRIDGE

French 7.5-mm Ball Cartridge

The French 7.5-mm ball cartridge is the most common of the wide variety of types of cartridges which are currently standard in the French Armed Forces. Reading the label on the box is the only positive method of field identification of specific rounds because of the many variations in identifying features.

7.5-mm cartridges are packed 15 rounds per cardboard box, with varying numbers of boxes in different-sized cases.

CHARACTERISTICS

Using weapons	MAS 36, M49, & M49/56 rifles; Model 1924/29 light machinegun; Models 31 and AA 52 machineguns
Overall length	2.99 in
Case material	Brass or steel
Muzzle velocity	810 m/sec
Max horizontal range	3000 m
Common factory designators	SF, TE, MI, VE, BS
Bullet type	Ball
Bullet weight	140 gr
Identifying markings	None

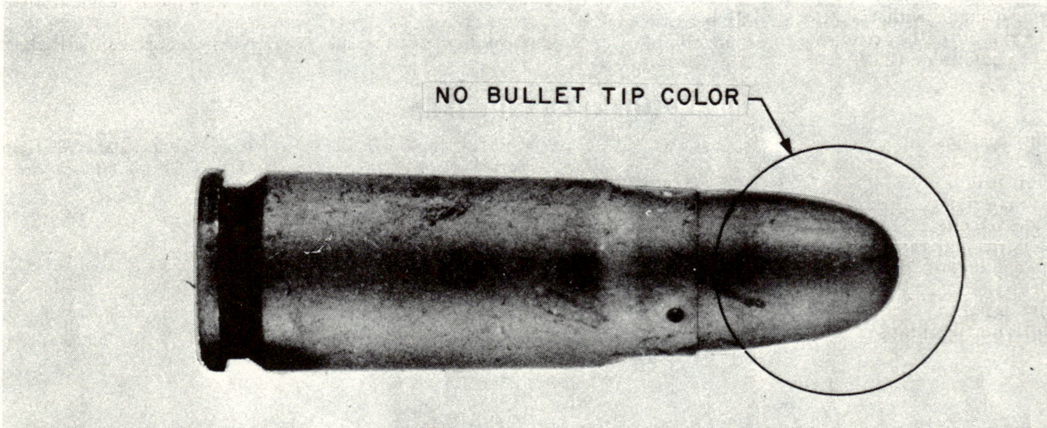

CHINESE COMMUNIST 7.62-mm TYPE 50 BALL CARTRIDGE

Chinese Communist 7.62-mm Type 50 Ball Cartridge

The Chinese Communist 7.62-mm Type 50 ball cartridge is copied from the Soviet 7.62-mm Type P cartridge. In Communist China, only ball cartridges are manufactured in this caliber, but the Soviet Type P cartridge was also made with tracer and API bullets.

Type 50 ball cartridges are packed 70 to a cardboard carton, 18 cartons in a galvanized metal container, and 2 containers (2520 rounds) in a wooden box.

CHARACTERISTICS

Using weapons	Types 51 and 54 pistols; Types 50 and 54 submachine-guns
Overall length	1.36 in
Case material	Brass
Muzzle velocity	420 to 500 m/sec
Max horizontal range	1800 m
Common factory designators	121, 11
Bullet type	Ball
Bullet weight	86 gr
Identifying markings	None

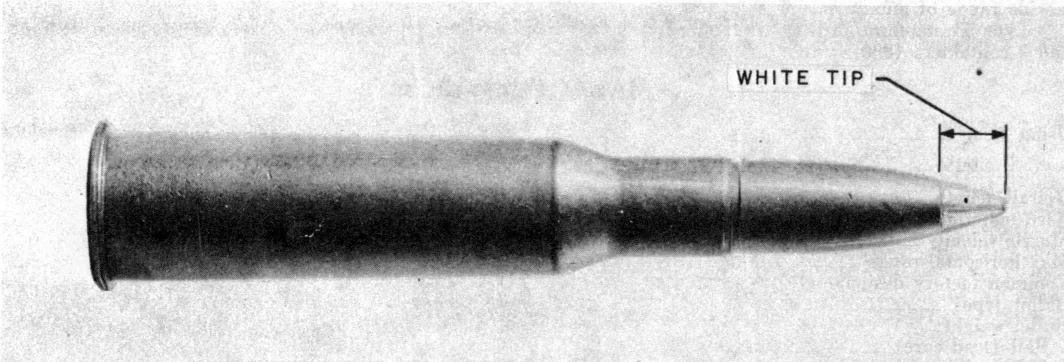

CHINESE COMMUNIST 7.62-mm TYPE 53 CARTRIDGE
(Ball--Mild Steel Core)

Chinese Communist 7.62-mm Type 53 Cartridges

The Chinese Communist 7.62-mm Type 53 cartridges are essentially copied from the Soviet 7.62-mm rimmed cartridges. The wide range in barrel lengths of the rifles and machineguns which fire these cartridges result in a wide range of muzzle velocities.

Type 53 ammunition is packed 20 rounds in a paper wrapper, 22 wrappers in a galvanized metal container, and 2 containers (880 rounds) in a wooden box.

CHARACTERISTICS

Using weapons	Type 53 carbines and light and heavy machineguns; Type 58 company machinegun; Type 57 heavy machinegun; and all Mosin Nagant rifles and carbines
Overall length	3.01 in
Case material	Brass
Muzzle velocity	805 to 865 m/sec
Max horizontal range	3100 m
Common factory designators	31, 71, 81
Bullet types	Ball (lead core), ball (mild steel core), and API
Bullet weight:	
Ball (lead core)	149 gr
Ball (mild steel core)	147 gr
API	155 gr
Identifying markings:	
Ball (lead core)	None
Ball (mild steel core)	White bullet tip
API	Black & red bullet tip

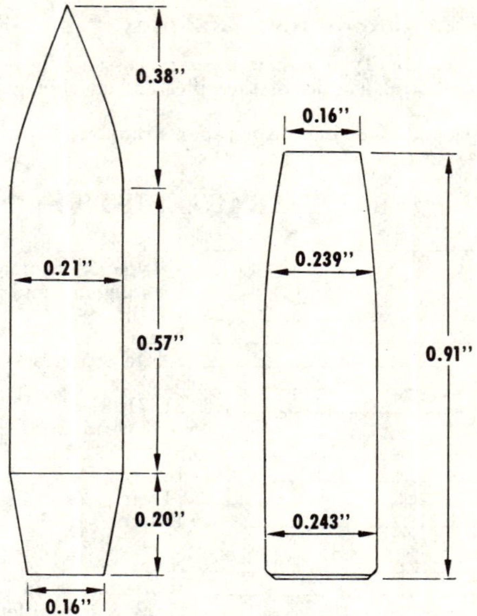

CHINESE COMMUNIST 7.62-mm TYPE 53 BULLET CORES

Ball (right) and API (left)

Chinese Communist 7.62-mm Type 53 Bullet Cores

Chinese Communist 7.62-mm Type 53 bullet cores may be recovered from various structures in a more-or-less intact condition. The two types manufactured by the Chinese Communists are illustrated. Aside from the difference in shapes, the ball bullet's mild steel core can be cut with a file while the API bullet's hard steel core cannot.

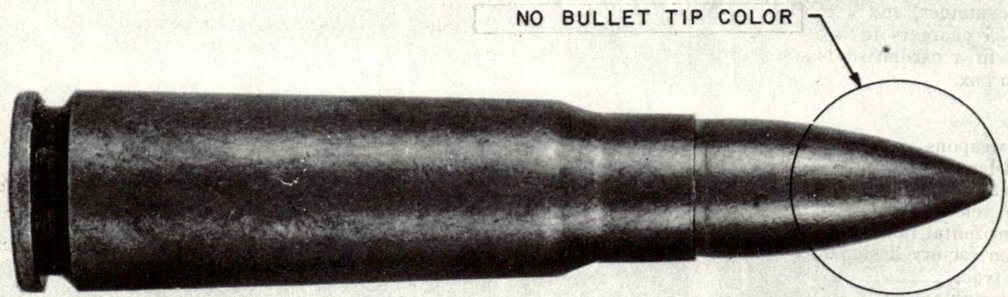

CHINESE COMMUNIST 7.62-mm TYPE 56 CARTRIDGE
(Ball)

Chinese Communist 7.62-mm Type 56 Cartridges

The Chinese Communist 7.62-mm Type 56 cartridges are copied from the Soviet 7.62-mm M43 cartridges. Types of specific rounds are identified by the color of the bullet tip (or absence of a colored bullet tip). The black paint in the extractor groove cut into the base of the cartridge case is not an identifying marking. Type 56 cartridges can be fired in a variety of weapons.

Type 56 cartridges are packed in three ways: 20 rounds in a paper wrapper, 36 wrappers in a galvanized metal container, and 2 containers (1440 rounds) in a wooden box; 10 rounds in a metal charger (stripper clip), 55 chargers in a galvanized metal container, and 2 containers (1100 rounds) in a wooden box; and 20 rounds in a cardboard box, 33 boxes in a galvanized metal container, and 2 containers (1320 rounds) in a wooden box.

CHARACTERISTICS

Using weapons	Type 56 carbines, assault rifles, and light machineguns
Overall length	2.18 in
Case material	Gilding-metal-clad steel
Muzzle velocity	710 to 745 m/sec
Max horizontal range	2500 m
Common factory designators	31, 71, 81
Bullet types	Ball, tracer, and API
Bullet weights:	
Ball	123 gr
Tracer	117 gr
API	117 gr
Identifying markings:	
Ball	None
Tracer	Green bullet tip
API	Black & red bullet tip

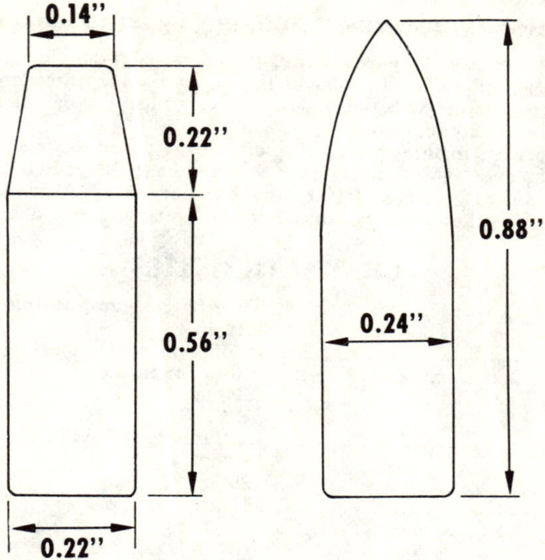

CHINESE COMMUNIST 7.62-mm TYPE 56 BULLET CORES

Ball (left) and API (right)

Chinese Communist 7.62-mm Type 56 Bullet Cores

Chinese Communist 7.62-mm Type 56 bullet cores, like the Type 53's described on page 9, are manufactured in two different types. Both are illustrated on the opposite page. The ball bullet's mild steel core can be easily cut with a file; the API bullet's hard steel core cannot.

CHINESE COMMUNIST 7.92 -mm BALL CARTRIDGE

Chinese Communist 7.92-mm Ball Cartridge

Chinese Communist 7.92-mm ball cartridges have been manufactured in Communist China for many years, but have never officially been given a type number. Cartridges with factory designators other than those listed below are likely to be encountered. 7.92-mm cartridges are packed 20 rounds in a paper wrapper and 26 wrappers (520 rounds) in a wooden box.

Recently, the Viet Cong have begun to use 7.92-mm ammunition manufactured in Czechoslovakia and Rumania. The Czechoslovak ball cartridges are packed 15 rounds in a cardboard box and 32 boxes (480 rounds) in a wooden case (gross weight, about 35 pounds). The Rumanian cartridges are packed 50 rounds in a cardboard box and 10 boxes (500 rounds) in a metal can (gross weight, about 35 pounds).

CHARACTERISTICS

Using weapons – Type 99 (converted Japanese) and M98–type (Mauser) rifles; MG 34, Bren Mk II, ZB–26 and ZB–30 light machineguns; and Type 24 heavy machinegun

Overall length:
- ChiCom ----- 3.16 in
- Czechoslovak – 3.16 in
- Rumanian --- 3.14 in

Case material -- Brass

Muzzle velocity:
- ChiCom ----- 825 m/sec
- Czechoslovak – 780 m/sec
- Rumanian --- 755 m/sec

Max horizontal range:
- ChiCom ----- 3100 m
- Czechoslovak – 4500 m
- Rumanian --- 5100 m

Common factory designators:
- ChiCom ----- 791, 451
- Czechoslovak – PS, O–★, ♂X
- Rumanian --- CMC

Bullet type:
- ChiCom ----- Ball (lead core)
- Czechoslovak – Ball (mild steel core)
- Rumanian --- Ball (lead core)

Bullet weight:
- ChiCom ----- 154 gr
- Czechoslovak – 178 gr
- Rumanian --- 198 gr

Identifying markings:
- ChiCom ----- None
- Czechoslovak – Black or navy primer annulus
- Rumanian --- Green primer annulus

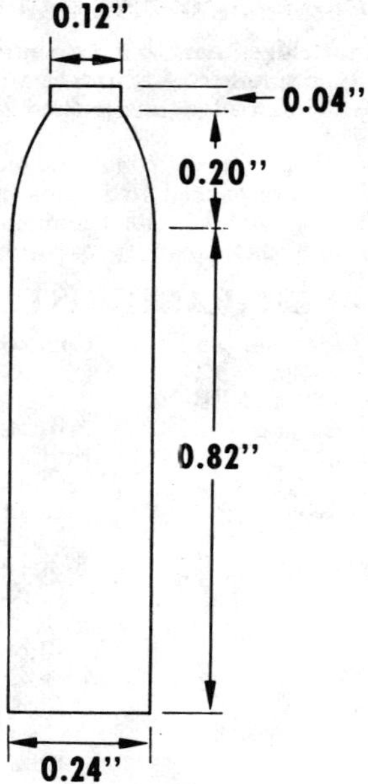

CZECHOSLOVAK 7.92-mm BALL BULLET CORE

Czechoslovak 7.92-mm Ball Bullet Core

The Czechoslovak 7.92-mm ball bullet core, because it is made of mild steel, may be recovered in more or less its original shape. (The Chinese Communist and Rumanian 7.92-mm bullet cores, made of lead, will be deformed beyond recognition when they strike a resistant structure.) The Czechoslovak bullet core can be easily cut with a file.

SOVIET 9-mm MAKAROV BALL CARTRIDGE

Soviet 9-mm Makarov Ball Cartridge

The Soviet 9-mm Makarov ball cartridge is not as powerful as the 9-mm Parabellum cartridge (p. 21); therefore it has lighter recoil in pistols and machine pistols and is less desirable as a submachine gun cartridge. Like most pistol cartridges, the Makarov is assembled only with ball-type bullets.

CHARACTERISTICS

Using weapons	Makarov pistol and Stechkin machine pistol
Overall length	0.97 in
Case material	Brass or steel
Muzzle velocity	315 to 340 m/sec
Max horizontal range	1400 m
Common factory designators	38, 270, 711
Bullet type	Ball
Bullet weight	94 gr
Identifying markings	None

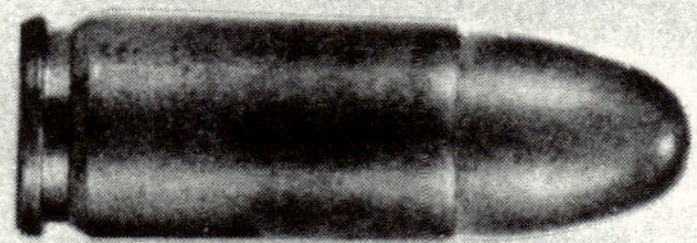

9-mm PARABELLUM BALL CARTRIDGE

9-mm Parabellum Ball Cartridge

The 9-mm Parabellum ball cartridge—probably the most popular and widely manufactured small arms round in the world—has been produced by almost every major nation, and a complete list of factory designators would be enormous. 9-mm Parabellum ammunition made in Australia, the United Kingdom, Canada, or France is most likely to be found in South Vietnam.

Packaging practices vary from country to country, but the most common pack is 50 rounds in a cardboard carton, 25 cartons in a can, and 2 cans (2500 rounds) in a wooden box.

CHARACTERISTICS

Using weapons	P–38, MAS 50, and Browning Hi-Power pistols; MAT–49, MP–40, F–1, Sten Marks 1, 2, 3, and 5, Chinese copy of Sten, and Chinese Type 37 submachine guns
Overall length	1.15 in
Case material	Brass
Muzzle velocity	380 to 400 m/sec
Max horizontal range	2000 m
Common factory designators	MF, MH, RG, TH, DI, DA, TE, TS, SF
Bullet type	Ball
Bullet weight	115 gr
Identifying markings	None

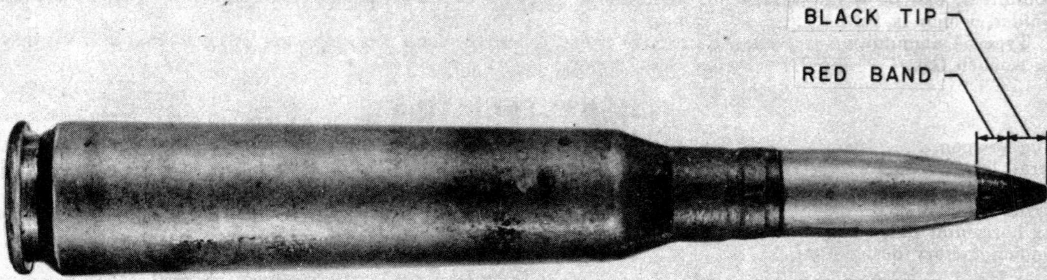

CHINESE COMMUNIST 12.7-mm TYPE 54 CARTRIDGE (API)

Chinese Communist 12.7-mm Type 54 Cartridges

The Chinese Communist 12.7-mm Type 54 cartridges have ballistics similar to U.S. caliber .50 cartridges but are not assembled with such a wide variety of bullets. Type 54 ammunition, copied from the Soviet 12.7-mm ammunition, can be readily distinguished from U.S. caliber .50 ammunition by the position of the cartridge shoulder, which is further forward, and by the case itself, which is considerably longer.

Type 54 ammunition is packed 85 rounds to a galvanized metal container and two containers (170 rounds) to a wooden box.

CHARACTERISTICS

Using weapon	Type 54 heavy machinegun
Overall length	5.78 in
Case material	Brass
Muzzle velocity	840 m/sec
Max horizontal range	7000 m
Common factory designators	41
Bullet types	API and API–T
Bullet weight:	
API	736 gr
API–T	681 gr
Identifying marks:	
API	Black & red bullet tip
API–T	Purple & red bullet tip

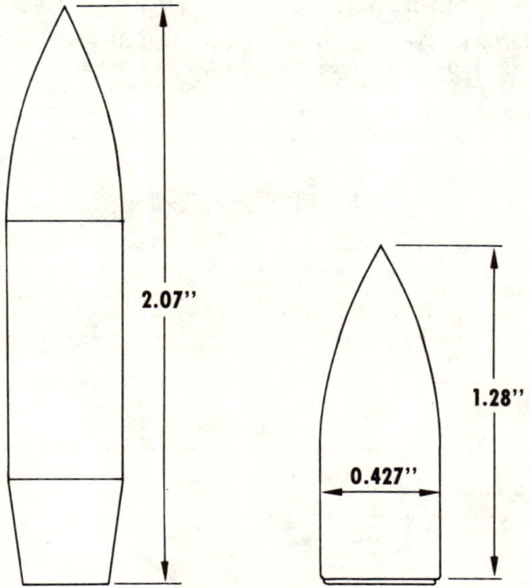

CHINESE COMMUNIST 12.7-mm TYPE 54 BULLET CORES
API (left) and API-T (right)

Chinese Communist 12.7-mm Type 54 Bullet Cores

Chinese Communist 12.7-mm Type 54 bullet cores are manufactured of hardened steel which cannot be cut with a file. They may be recovered more or less intact. The API bullet's longer core has a tendency to break across its middle (cylindrical portion) if it strikes a resistant structure at large angles of obliquity.

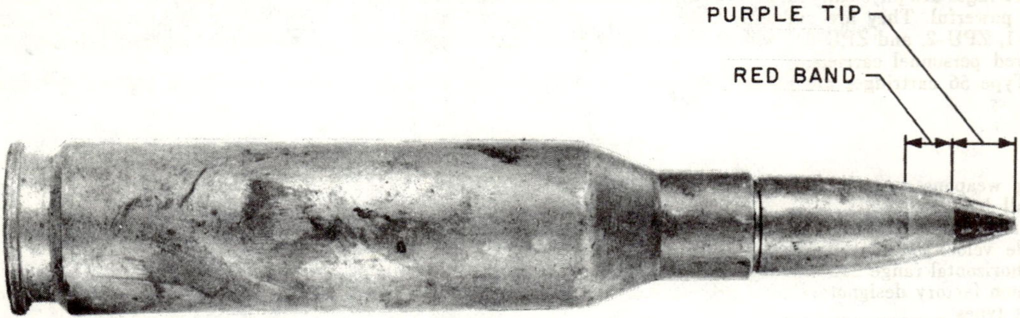

CHINESE COMMUNIST 14.5-mm TYPE 56 CARTRIDGE (API-T)

Chinese Communist 14.5-mm Type 56 Cartridges

The Chinese Communist 14.5-mm Type 56 cartridges are copied from the Soviet 14.5-mm ammunition. Type 56 cartridges are physically large, in comparison with most other small arms ammunition, and are correspondingly more powerful. They are currently fired only in the Type 56 heavy antiaircraft machineguns (Soviet Models ZPU-1, ZPU-2, and ZPU-4), which, although primarily antiaircraft weapons, are effective against vehicles and armored personnel carriers.

Type 56 cartridges are packed 42 rounds in a galvanized container and two containers (84 rounds) to a box.

CHARACTERISTICS

Using weapons	Type 56 heavy antiaircraft machineguns
Overall length	6.12 in
Case material	Brass
Muzzle velocity	1000 m/sec
Max horizontal range	7000 m
Common factory designators	41
Bullet types	API, API–T, and I–T
Bullet weight:	
API	990 gr
API–T	915 gr
I–T	921 gr
Identifying markings:	
API	Black & red bullet tip
API–T	Purple & red bullet tip
I–T	Red bullet tip

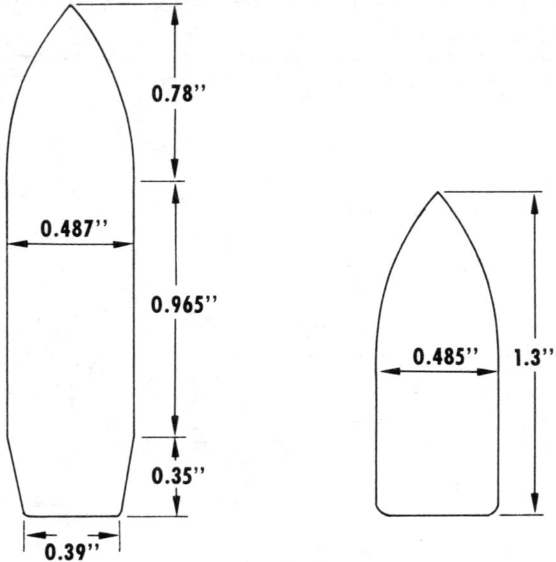

CHINESE COMMUNIST 14.5-mm TYPE 56 BULLET CORES

API (left) and API-T (right)

Chinese Communist 14.5-mm Type 56 Bullet Cores

Chinese Communist 14.5-mm Type 56 bullet cores of the API and API-T rounds are made of hardened steel which cannot be cut with a file. The longer core of the API bullet has a tendency to break across its middle (cylindrical portion) on impacting at large angles of obliquity.

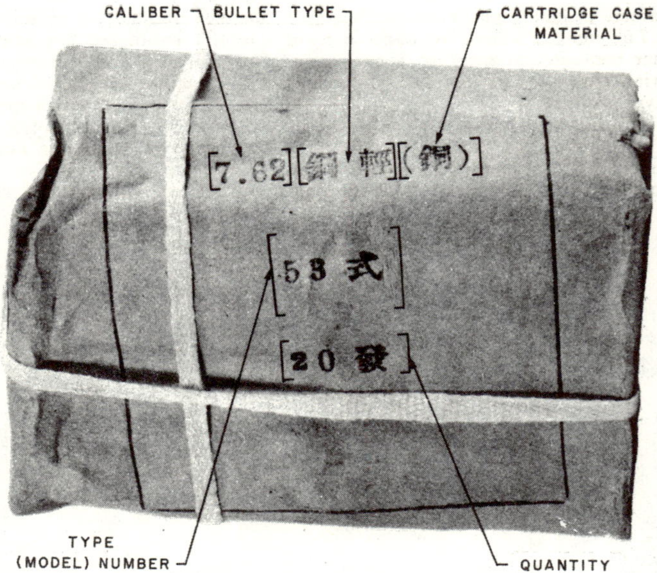

INNER CONTAINER (PAPER WRAPPING)

Inner Container (Paper Wrapping)

The inner container, made of paper or cardboard, is the innermost of three distinct containers in which Chinese Communist 7.62-mm cartridges are packed. Except for the cardboard box containing 7.62-mm Type 50 cartridges, each inner container is marked to show the nature of its contents, normally: Caliber, bullet type, cartridge case material, type (model) number, and quantity. The caliber and type number are shown in Arabic numerals, and the type number is always followed by the Chinese character which means "type." Bullet types and cartridge case materials are indicated by Chinese characters, which are printed on the inner wrapper in script either in full or abbreviated. See appendix I, part 1, for the script form of a number of common Chinese characters. The illustration shows the location and meaning of the characters found on a typical inner container. ***NOTE:*** An inner container may not carry all of the data listed above and shown in the illustration.

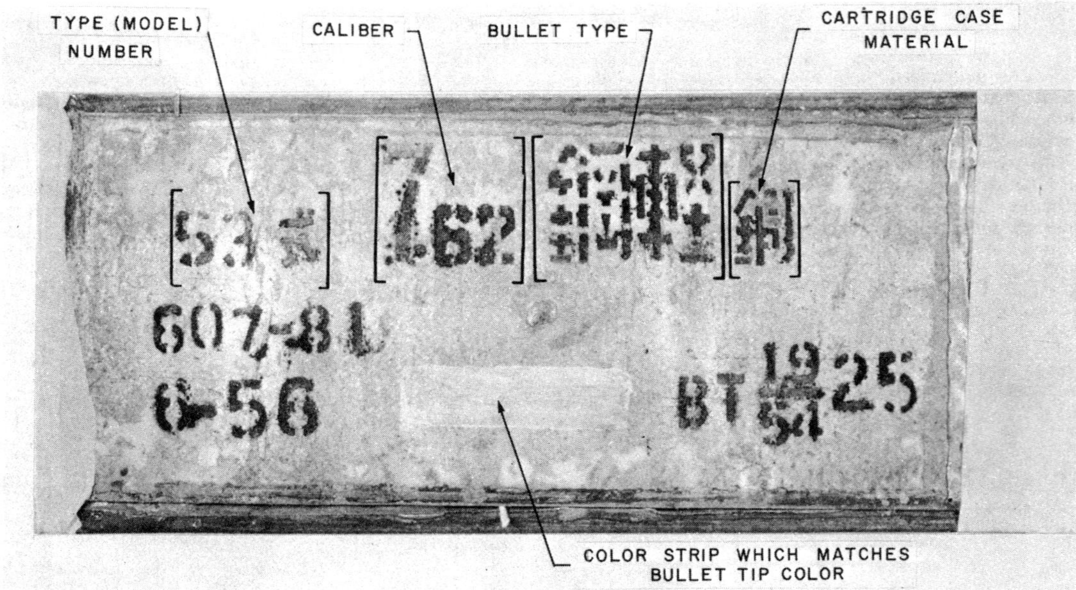

GALVANIZED METAL CAN

Galvanized Metal Can

The galvanized metal can is the second container in which Chinese Communist 7.62-mm cartridges are packed and the innermost container for the 12.7-mm and 14.5-mm cartridges. It in turn is packed in a wooden box. The galvanized metal can is marked to show the nature of its contents, normally: Caliber, bullet type, cartridge case material, type (model) number, and quantity.

The caliber and type number are shown in Arabic numerals, and the type number is always followed by the Chinese character which means "type." Bullet types and cartridge case materials are indicated by Chinese characters which are stenciled, sometimes in full and sometimes abbreviated, on the container. See appendix 1, part 2, for a list in stenciled form of those characters which either apply or are peculiar to small arms ammunition. *NOTE:* Each container does not always carry all of the markings listed above and shown in the illustration.

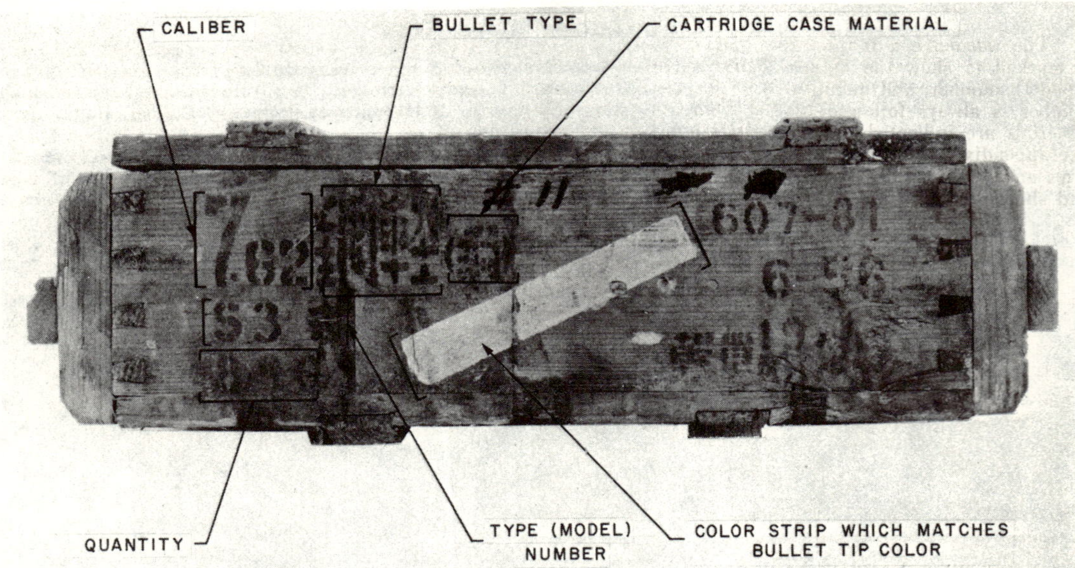
CALIBER
BULLET TYPE
CARTRIDGE CASE MATERIAL
QUANTITY
TYPE (MODEL) NUMBER
COLOR STRIP WHICH MATCHES BULLET TIP COLOR

WOODEN BOX

Wooden Box

The wooden box is the outermost container in which Chinese Communist cartridges are packed. The box is marked to show the nature of its contents, normally: Caliber, bullet type, cartridge case material, type (model) number, and quantity. The caliber and the type number are shown in Arabic numerals, and the type number is always followed by the Chinese character which means "type." Bullet types and cartridge case materials are indicated by Chinese characters which are stenciled on the box either in full or abbreviated. See appendix 1, part 2, for the stenciled form of those characters which either apply or are peculiar to small arms ammunitions. *NOTE:* Each wooden box container does not always carry all of the markings noted above and shown in the illustration.

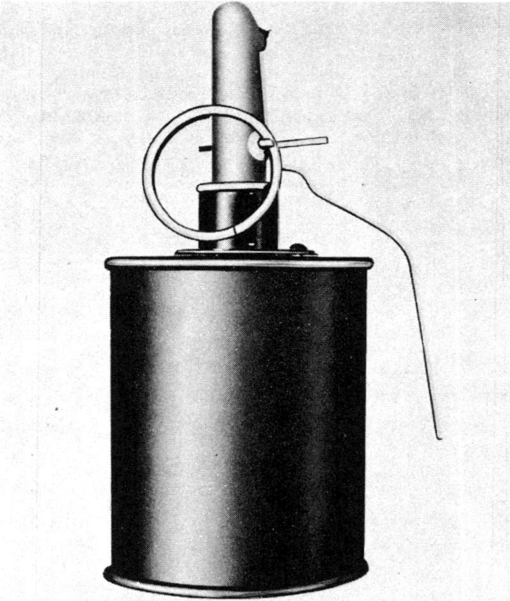

CHINESE COMMUNIST TYPE 42 OFFENSIVE /DEFENSIVE HAND GRENADE

Chinese Communist Type 42 Offensive/Defensive Hand Grenade

The Chinese Communist Type 42 offensive/defensive hand grenade, copied from the Soviet Model RG–42, has a body, a charge of pressed TNT, and a fuze. The outer part of the body is a cylindrical sheet steel can with a boss riveted to one end. A threaded hole through the boss allows the fuze to be assembled to the body. The inner part of the grenade body is a sheet steel fragmentation liner which has been scored in a checkerboard pattern. The fuze assembly incorporates a spring-loaded striker held in place by a safety lever. A detonator containing a 3- to 4-second delay element in fastened to the threaded portion of the fuze housing, in line with the striker.

Type 42 grenades are shipped with brown plastic plugs screwed into the fuze wells in the top of the grenade bodies. Arming and operation of these grenades are the same as for the Type 1 defensive hand grenade (p. 41). Twenty Type 42 grenades with fuzes are packed in a wooden box, having a gross weight of about 33 pounds.

CHARACTERISTICS

Overall length	5.0 in
Total weight	0.79 lb
Max diameter	2.3 in
Identifying markings	Type number stenciled on bottom of grenade body
Filler material	Pressed TNT
Filler weight	3.9 oz.
Fragmentation radius	15 m

CHINESE COMMUNIST STICK-TYPE DEFENSIVE HAND GRENADE

Chinese Communist Stick-Type Defensive Hand Grenade

The Chinese Communist stick-type defensive hand grenade is found in a variety of head sizes and shapes; some are scored or serrated and some are not. Explosive fillers include picric acid, mixtures of TNT or nitroglycerin with potassium nitrate or sawdust, and schneiderite. All stick-type defensive hand grenades function in the same manner. The cord of a pull-friction fuze is located underneath the cap at the end of the throwing handle. A tug on this cord ignites a delay element; 2.5 to 6 seconds later the detonator explodes the main charge. Stick-type hand grenades are dangerous; they should be handled only when necessary and then only with caution.

These grenades are most commonly packed 20 fuzed grenades to a wooden box.

CHARACTERISTICS

Overall length	8.0 to 9.7 in
Total weight	1.16 to 1.22 lb
Max diameter	1.7 to 2.2 in
Identifying markings	None
Filler material	Varies widely
Filler weight	0.9 to 2.2 oz
Fragmentation radius	10 m

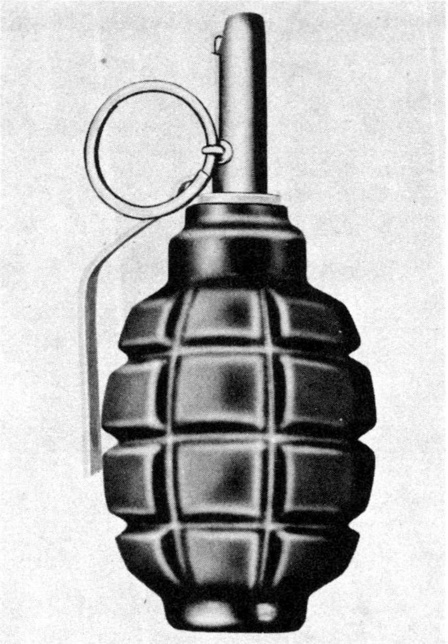

CHINESE COMMUNIST TYPE 1 DEFENSIVE HAND GRENADE

Chinese Communist Type 1 Defensive Hand Grenade

The Chinese Communist Type 1 defensive hand grenade, copied from the Soviet Model F-1, has a serrated, cast iron body containing a charge of cast TNT and a fuze. A brown plastic plug threaded into the fuze well in the top of the body prevents the entry of moisture and foreign matter. The fuze consists of a delay detonator, housing, spring-loaded striker, safety lever, and a safety ring attached to a cotter pin.

To arm and use this grenade, first unscrew and discard the brown plastic plug. Second, after checking to be sure that the fuze well is not obstructed, screw a fuze into the fuze well. Third, grasping both the grenade and fuze safety lever with one hand, pull-out the safety ring and cotter pin. As the grenade is thrown, the compressed striker spring forces the striker down into the primer and the delay starts burning. After 3 to 4 seconds, the delay initiates the detonator, which then sets off the charge of TNT.

Twenty Type 1 grenades with fuzes are packed in a wooden box having a gross weight of a little more than 43 pounds.

CHARACTERISTICS

Overall length	4.9 in
Total weight	1.28 lb
Max diameter	2.2 in
Identifying markings	Type number stenciled on bottom of grenade body
Filler material	Cast TNT
Filler weight	1.94 oz
Fragmentation radius	15 m

CHINESE COMMUNIST TYPE 59 DEFENSIVE HAND GRENADE

Chinese Communist Type 59 Defensive Hand Grenade

The Chinese Communist Type 59 defensive hand grenade, copied from the Soviet RGD-5, has an egg-shaped, sheet metal body which incloses a nonserrated fragmentation liner. The fuze is a conventional cocked-striker mechanism which functions in the following manner: First the safety ring and pin are withdrawn. Then when the safety lever is released, the compressed striker spring drives the striker into a primer. The primer ignites a 3- to 4-second delay element which in turn initiates a detonator to explode the main filler charge of TNT.

Unfuzed Type 59 grenades are packed 20 to a wooden box, together with 20 fuzes. The box and contents weigh slightly under 31 pounds.

CHARACTERISTICS

Overall length	4.5 in
Total weight	0.68 lb
Max diameter	2.1 in
Identifying markings	Type number stenciled on bottom of grenade body
Filler material	TNT
Filler weight	3.9 oz
Fragmentation radius	20 m

SOVIET RKG-3 HEAT HAND GRENADE FAMILY

Soviet RKG-3 HEAT Hand Grenade Family

The Soviet RKG–3 HEAT hand grenade family comprises the RKG–3, RKG–3M, and RKG–3T, all similar in appearance. Each has a handle which contains a parachute-shaped stabilizer, a middle section which contains a fuze assembly, and a head which contains a shaped charge. When any of these grenades is thrown, a spring inside the handle ejects the stabilizer, which keeps the shaped charge pointed in the direction the grenade was thrown. When the grenade strikes the target or other hard object, inertia causes a firing pin to strike a primer, which, in turn initiates a detonator to explode the main charge.

CHARACTERISTICS

Overall length	14.2 in
Total weight	2.14 lb
Max diameter	2.2 in
Identifying markings	Model number stenciled on front face of charge container
Filler material	TNT/RDX mixture
Filler weight	20 oz
Penetration	Approx 7 in

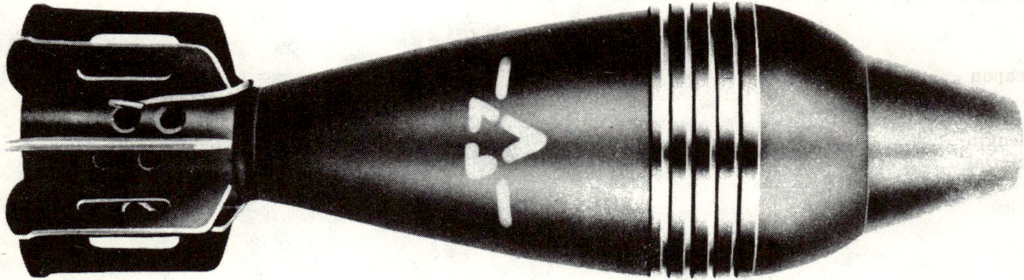

CHINESE COMMUNIST 60-mm HE CARTRIDGE

Chinese Communist 60-mm HE Cartridge

The Chinese Communist 60-mm HE cartridge is a mortar projectile generally similar to the U.S. 60-mm mortar projectiles of the M49 series. This cartridge, a fin-stabilized design, has an HE-loaded projectile body, point detonating fuze, fin assembly, shotgun-shell-type ignition cartridge, and propellent increment charges which can be adjusted for zone fire. Increments, in rectangular cloth bags, are inserted between the fins under the spring wire retainers of the fin assembly.

Cartridges are packed 10 rounds in a wooden box.

CHARACTERISTICS

Using weapon	Type 31 mortar
Caliber	60 mm
Weight	3.3 lb
Overall length	8.9 in
Body material	Cast iron

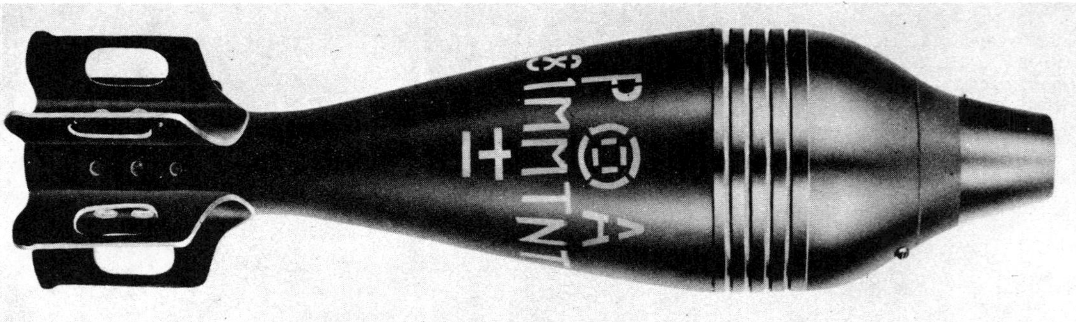

CHINESE COMMUNIST 81-mm HE CARTRIDGE

Chinese Communist 81-mm HE Cartridge

The Chinese Communist 81-mm HE cartridge is a conventional-type mortar round. It has a point detonating fuze, fuze adapter, steel projectile, fin assembly, and propellent increment charges. The increment charges, in rectangular cloth bags, are held between the fins by wire retainers.

Four rounds, with fuzes and increment charges in separate containers, are packed in a wooden box.

CHARACTERISTICS

Using weapon	81-mm mortar (copy of U.S. 81 mm)
Caliber	81 mm
Weight	7.2 lb
Overall length	12.8 in
Body material	Steel

CHINESE COMMUNIST 82-mm HE CARTRIDGE

Chinese Communist 82-mm HE Cartridge

The Chinese Communist 82-mm HE cartridge is of conventional design. A teardrop-shaped mortar cartridge, it has a point detonating fuze, steel body, fin assembly, and propellent increment charges. The heavy paper ignition cartridge is inserted over a threaded brass primer. The propellent increments may be inserted through the slots in the fins and held in place by spring clips.

The rounds are packed three or four per wooden box with fuzes and increments in a separate container within the box.

CHARACTERISTICS

Using weapons	Types 20 and 53 mortars
Caliber	82 mm
Weight	8.6 lb
Overall length	12.8 in
Body material	Steel

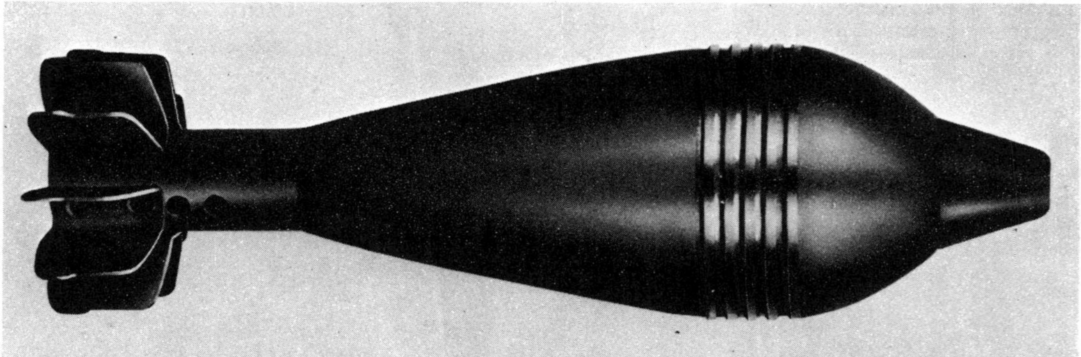

SOVIET 82-mm MODEL VO-832D FRAG CARTRIDGE

Soviet 82-mm Model VO–832D Frag Cartridge

The Soviet 82-mm Model VO–832D frag mortar cartridge has an M–5 point detonating fuze, cast iron projectile, fin assembly, propellent increment charges, and a shotgun-shell-type ignition cartridge. The split ring propellent increments are inclosed in a cotton cloth bag.

Ten fuzed rounds with increments are packed in a wooden box with a gross weight of about 106 pounds. This cartridge may also be packed three rounds in a metal box.

CHARACTERISTICS

Using weapon	M1937 mortar
Caliber	82 mm
Weight	6.9 lb
Overall length	12.3 in
Body material	Cast iron

CHINESE COMMUNIST 120-mm HE CARTRIDGE (SHORT BODY)

Chinese Communist 120-mm HE Cartridge (Short Body)

The Chinese Communist 120-mm HE cartridge (short body) is constructed of cast iron. The white silk increment charge bags are placed in the elongated holes in the fin assembly or tied to the fin assembly with string. This stubby mortar round has a cardboard ignition cartridge and a threaded primer. The mouth of the projectile is reduced by a steel adapter, which accepts the brass point detonating fuze and steel booster cup.

Two cartridges are packed in a wooden box, together with a zinc-coated tin box containing increments and fuzes for firing two rounds.

CHARACTERISTICS

Using weapon	Type 33 mortar
Caliber	120 mm
Weight	28.9 lb
Overall length	19.2 in
Body material	Cast iron

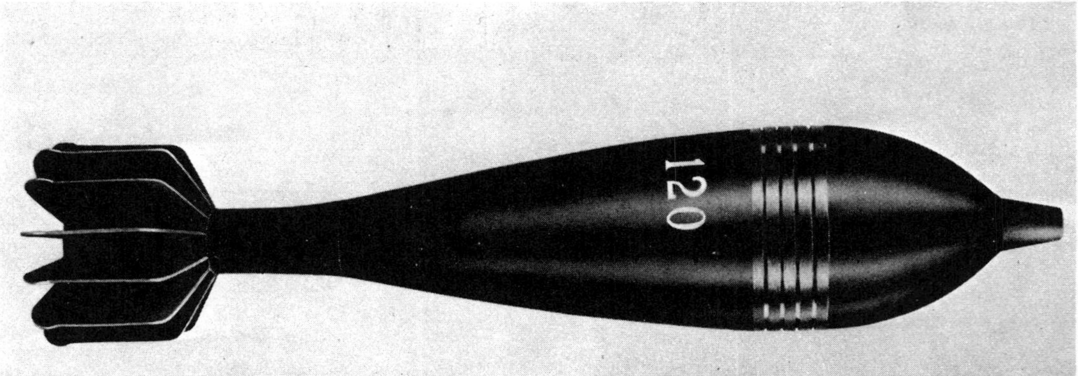

CHINESE COMMUNIST 120-mm HE CARTRIDGE (LONG BODY)

Chinese Communist 120-mm HE Cartridge (Long Body)

The Chinese Communist 120-mm HE mortar cartridge (long body) has a point detonating fuze, cast iron fuze adapter, steel projectile body, and fin assembly. Increment charges, which are tied around the fin assembly shaft, are initiated by a threaded brass primer and an ignition cartridge.

One round and an accessory box (containing a fuze, primer, ignition cartridge, and increments) are packed in a wooden box.

CHARACTERISTICS

Using weapon	Type 33 mortar
Caliber	120 mm
Weight	28 lb
Overall length	23.5 in
Body material	Steel

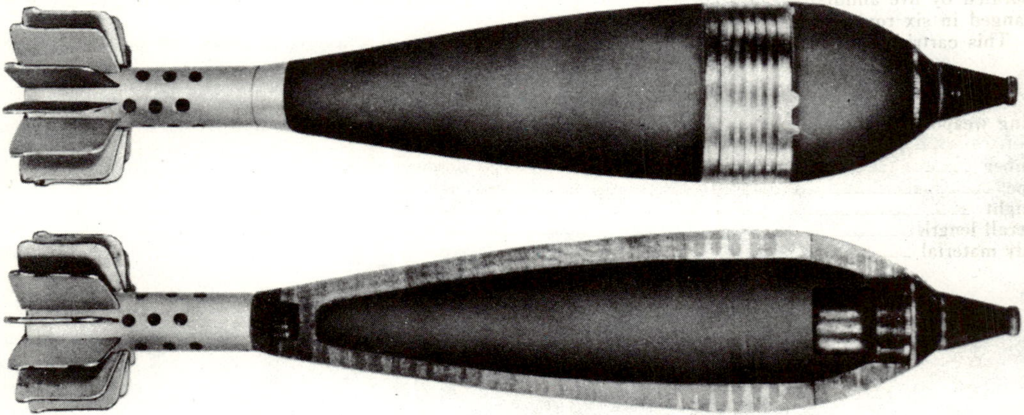

SOVIET 120-mm MODEL VOF-843A FRAG-HE CARTRIDGE

Soviet 120-mm Model VOF–843A Frag–HE Cartridge

The Soviet 120-mm Model VOF–843A frag–HE mortar cartridge is of teardrop design. Six gas-check bands separated by five annular grooves constitute the bourrelet. The fin assembly has 12 fins and 18 gas-escape ports arranged in six rows of three.

This cartridge is packed two rounds per wooden box with a gross weight of about 100 pounds.

CHARACTERISTICS

Using weapons	M1938 and M1943 regimental mortars; ChiCom Type 53 mortar
Caliber	120 mm
Type	Frag HE
Weight	36.3 lb
Overall length	26.3 in
Body material	Cast iron

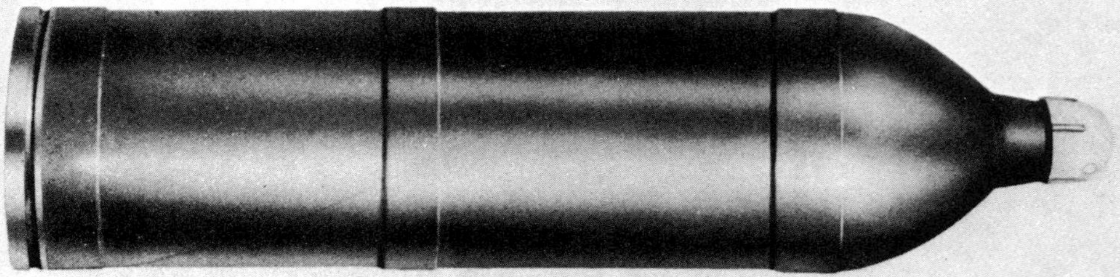

CHINESE COMMUNIST 87-mm TYPE 135 HEAT ROCKET

Chinese Communist 87-mm Type 135 HEAT Rocket

The Chinese Communist 87-mm Type 135 HEAT spin-stabilized rocket has a rocket motor with eight canted nozzles, a head with a base detonating element, and a noze fuze with spit-back action (PIBD). The head and motor body appear to be machined from steel tubing or bar stock and the nose is cast iron.

The rocket is painted either olive drab with markings stenciled in yellow, or black with markings stenciled in white.

CHARACTERISTICS

Using weapon	Type 51 rocket launcher
Caliber	87 mm
Weight	11.7 lb
Overall length	14.7 in
Type of stabilization	Spin
Armor penetration	3.75 in at 0°

CHINESE COMMUNIST 87-mm TYPE 241 HEAT ROCKET

Chinese Communist 87-mm Type 241 HEAT Rocket

The Chinese Communist 87-mm Type 241 HEAT rocket resembles the U.S. 3.5-inch M28A2 rocket in general appearance. The fin-stabilized rocket has a motor body, nozzle, fin shroud assembly, trap, propellent spacer, adapter with wooden plug, igniter assembly, fuze, and head. The motor is roughly machined inside and out. The fin shroud assembly surrounds the nozzle and is joined to it by four struts, which are fixed by spot welding. The adapter acts as a motor closure and joins the head to the motor. The head section consists of an ogive, nose adapter, head body, shaped charge cone, flash tube, detonator booster, and main charge. The fuze is a point-initiating, spit-back type (PIBD).

The Type 241 rocket is painted either olive drab with markings stenciled in yellow, or black with white-stenciled markings.

CHARACTERISTICS

Using weapon	Type 51 rocket launcher (copy of U.S. 3.5 inch)
Caliber	87 mm
Weight	8.4 lb
Overall length	19.5 in
Type of stabilization	Fin
Armor penetration	4 in at 0°

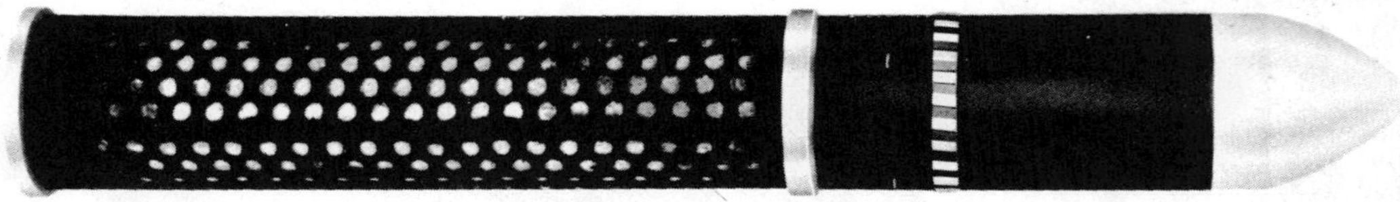

CHINESE COMMUNIST 57-mm HEAT CARTRIDGE

Chinese Communist 57-mm HEAT Cartridge

The Chinese Communist 57-mm HEAT cartridge externally resembles U.S. 57-mm recoilless ammunition. The fuze, made of aluminum alloy, is unusually heavy, for it serves as the ogive of the projectile and its lower part is machined to form the bourrelet. The projectile's rotating band is preengraved. The perforated cartridge case, which is welded to the base, is manufactured from two halves of strip steel welded longitudinally. Four equally spaced slots and stab crimps are at the mouth of the cartridge case.

Each round is packed in a hermetically sealed metal container, and five containers are packed per wooden box.

CHARACTERISTICS

Using weapon	Type 36 recoilless rifle (copy of U.S. M18)
Caliber	57 mm
Weight	5.7 lb
Overall length	18 in
Projectile weight	2.7 lb
Color	Black
Case length	11.9 in
Armor penetration	3 in at 0°

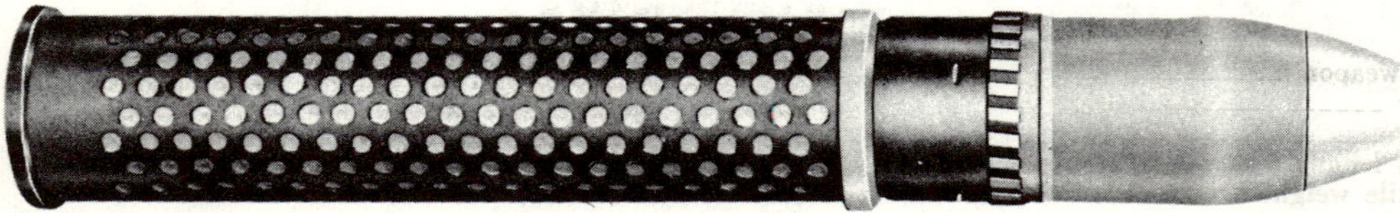

CHINESE COMMUNIST 57-mm HE CARTRIDGE

Chinese Communist 57-mm HE Cartridge

The Chinese Communist 57-mm HE cartridge has a projectile with point detonating fuze and a primed perforated cartridge case containing the propellent charge. This cartridge is, in general, copied from U.S. 57-mm recoilless ammunition. The projectile has a preengraved rotating band. The perforated cartridge, which is welded to the base, is manufactured from two halves of strip steel welded longitudinally. Four equally spaced slots and stab crimps are at the mouth of the cartridge case.

Each round is packed in a hermetically sealed metal container, and five containers are packed in a wooden box.

CHARACTERISTICS

Using weapon	Type 36 recoilless rifle (copy of U.S. M18)
Caliber	57 mm
Weight	5.7 lb
Overall length	17.5 in
Projectile weight	2.8 lb
Color	Yellow
Case length	11.9 in

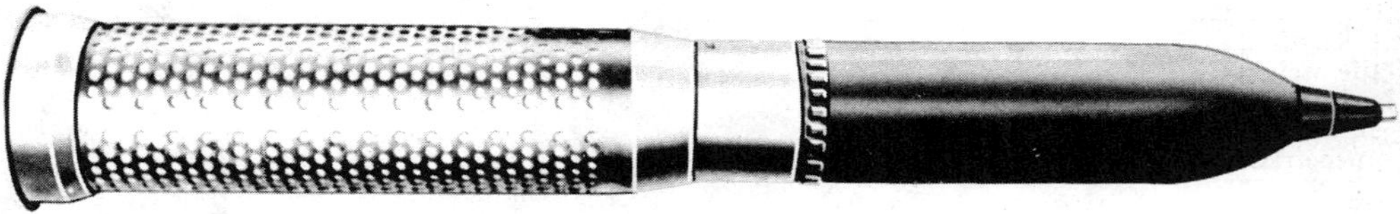

CHINESE COMMUNIST 75-mm HEAT CARTRIDGE

Chinese Communist 75-mm HEAT Cartridge

The Chinese Communist 75-mm HEAT cartridge resembles, in general, the U.S. 75-mm recoilless rifle cartridge. The projectile is of different construction, however, and uses a point-initiating, base-detonating fuze. The projectile has a preengraved rotating band, and the cartridge case is multiperforated.

The round, completely assembled, is stored and shipped in an individual hermetically sealed, reusable metal container. The container has a carrying strap and is painted olive drab with white stenciled markings.

CHARACTERISTICS

Using weapons	Types 52 (copy of U.S. M20) and 56 recoilless rifles
Caliber	75 mm
Weight	21 lb
Overall length	29 in
Projectile weight	13 lb
Case length	16 in
Color	Gray
Armor penetration	6 in at 0°

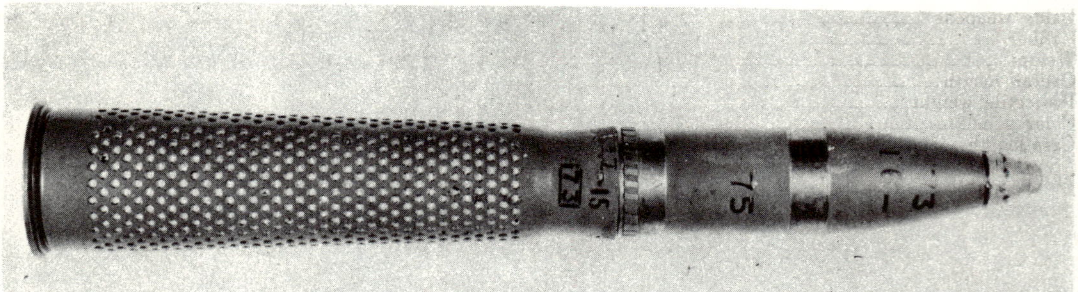

CHINESE COMMUNIST 75-mm HE CARTRIDGE

Chinese Communist 75-mm HE Cartridge

The Chinese Communist 75-mm HE cartridge is similar in construction to U.S. 75-mm recoilless ammunition. This recoilless rifle cartridge is distinguished by a bottle-necked perforated cartridge case and a preengraved rotating band on the projectile. The case is crimped to the projectile by four evenly spaced indentations.

CHARACTERISTICS

Using weapons	Types 52 (copy of U.S. M20) and 56 recoilless rifles
Caliber	75 mm
Weight	20.4 lb
Overall length	27 in
Projectile weight	12 lb
Color	Gray
Case length	15.8 in

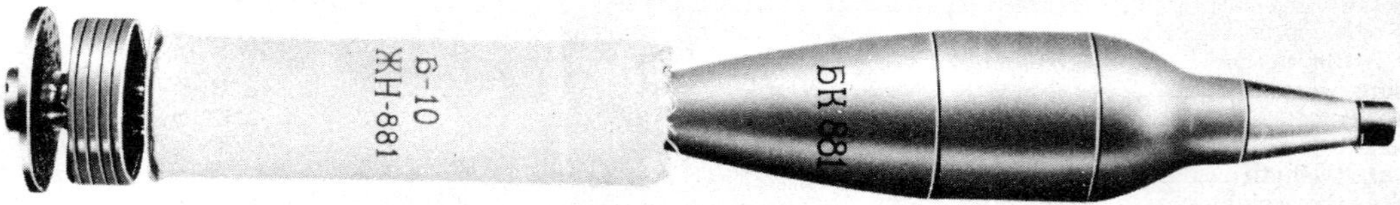

SOVIET 82-mm MODEL BK-881 HEAT CARTRIDGE

Soviet 82-mm Model BK–881 HEAT Cartridge

The Soviet 82-mm Model BK–881 HEAT cartridge consists basically of a steel projectile body, a nose fuze, a stabilizer assembly, and a caseless propellent charge. The projectile body, which houses a shaped charge, has a rather thick wall for a HEAT-type projectile. The explosive filler is 94% RDX and 6% wax, and the cartridge uses the GK–2 fuze. The stabilizer assembly is threaded to the base of the projectile body. The propellent charge is a strip-type, double-base propellant in a combustible cloth bag and is fitted securely around the stabilizer shaft.

Each Model BK–881 cartridge is packed in a tubular steel, hermetically sealed container, and four containers are packed in a wooden box.

CHARACTERISTICS

Using weapon	B–10 recoilless gun
Caliber	82 mm
Weight	10.8 lb
Overall length	27 in
Projectile weight	8 lb
Color	Olive drab
Armor penetration	11.8 in at 0°

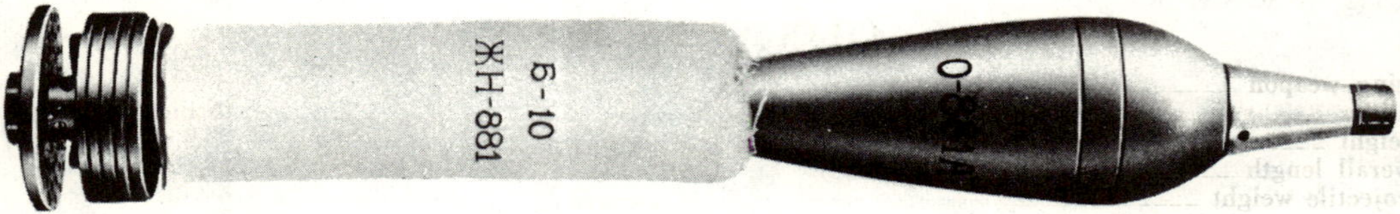

SOVIET 82-mm MODEL 0-881A FRAG CARTRIDGE

Soviet 82-mm Model O–881A Frag Cartridge

The Soviet 82-mm Model O–881A frag cartridge has a projectile body, a nose fuze, a stabilizer assembly, and a caseless propellent charge. The cast iron, teardrop-shaped projectile is filled with an explosive charge of TNT and dinitronaphthalene. The GK–2 point detonating fuze is threaded to the nose of the projectile; a hollow shaft serving a stabilizer assembly is attached at the base. The propellent charge, of double base powder is contained in a combustible cloth bag, and is fitted tightly around the stabilizer shaft.

Each cartridge is packed in a tubular steel, hermetically sealed container, and four containers are packed in a wooden box.

CHARACTERISTICS

Using weapon	B–10 recoilless gun
Caliber	82 mm
Weight	10.8 lb
Overall length	24 in
Projectile weight	8.6 lb
Color	Gray

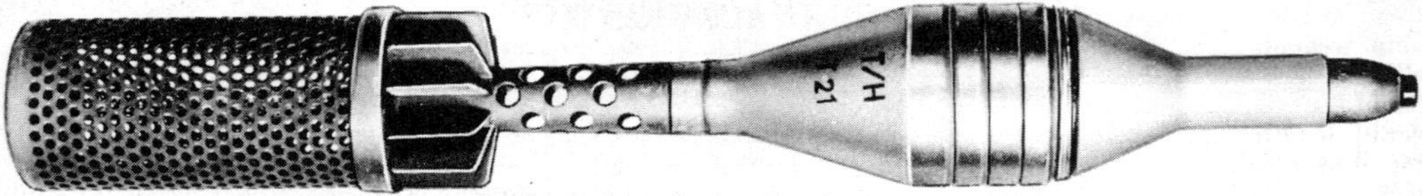

CZECHOSLOVAK 82-mm MODEL T-21 HEAT CARTRIDGE

Czechoslovak 82-mm Model T–21 HEAT Cartridge

The Czechoslovak 82-mm Model T-21 HEAT cartridge has fin-stabilized HEAT projectile, a point-initiating, base-detonating fuze, and a perforated steel cartridge case. The thin propellent wafers are inserted into the front of the cartridge case and fit around the electric primer body located within the case. A threaded steel plate closes the mouth of the case and holds the propellant in place. The plate also has a steel spigot-type protrusion for attaching the projectile to the case. The Z–21 fuze has a steel body and a plastic striker head.

Six unfuzed rounds are packed in a wooden box (the fuzes are packed in a separate box).

CHARACTERISTICS

Using weapon	T–21 (Tarasnice) recoilless gun
Caliber	82 mm
Weight	7.9 lb
Overall length	24.7 in
Projectile weight	4.7 lb
Color	Dark green with red stripe
Case length	6.7 in
Armor penetration	9 in at 0°

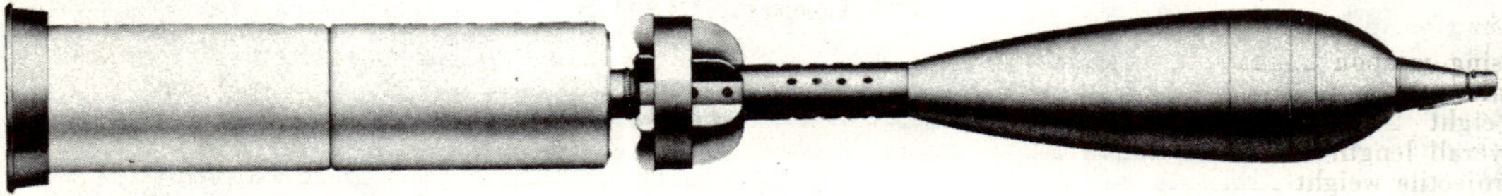

SOVIET 107-mm MODEL BK-883 HEAT CARTRIDGE

Soviet 107-mm Model BK–883 HEAT Cartridge

The Soviet 107-mm Model BK–883 HEAT cartridge has a fin-stabilized steel projectile, a GK–2 fuze, and a separate but attachable caseless propellent charge. The projectile contains a shaped charge of RDX and aluminum, and a tracer unit is located within the shaft of the fin assembly. The propellent charge, surrounding a flash tube and assembled to a metal base, consists of two increment charges of stick propellant each wrapped in a combustible linen cover. Before the projectile is loaded into the weapon, the propellent charge is attached to the base of the fin assembly by means of a snap-on fastener.

Two cartridges, with propellent charges in separate airtight containers, are packed in a wooden box.

CHARACTERISTICS

Using weapon	B–11 recoilless gun
Caliber	107 mm
Weight	27.6 lb
Overall length	43.4 in
Projectile weight	16.5 lb
Color	Olive drab
Armor penetration	15 in at 0°

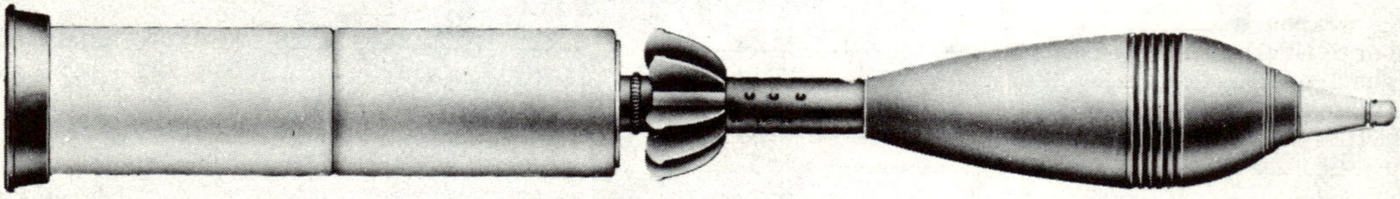

SOVIET 107-mm MODEL OF-883A FRAG-HE CARTRIDGE

Soviet 107-mm Model OF–883A Frag–HE Cartridge

The Soviet 107-mm Model OF–833A frag–HE cartridge has a fin-stabilized cast-iron projectile, to which the GK–2 point detonating fuze is assembled, and a separate but attachable caseless propellent charge. The projectile is filled with 80/20 amatol. The propellent charge, surrounding a flash tube and assembled to a metal base, consists of two increment charges of stick propellant each wrapped in a combustible linen cover. Before the projectile is loaded into the weapon, the propellent charge is attached to the base of the fin assembly by means of a snap-on fastener.

Two cartridges, with propellent charges in separate airtight containers, are packed in a wooden box.

CHARACTERISTICS

Using weapon	B–11 recoilless gun
Caliber	107 mm
Weight	29.8 lb
Overall length	38.3 in
Projectile weight	18.7 lb
Color	Olive drab

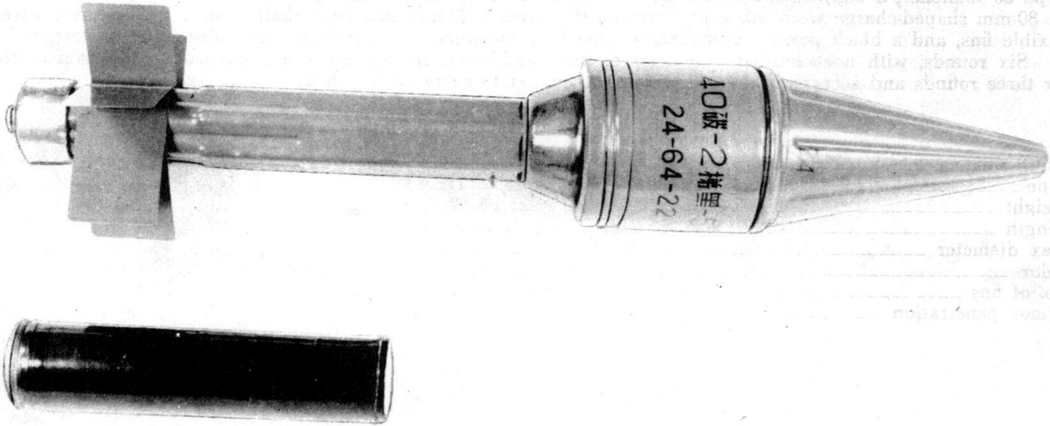

CHINESE COMMUNIST TYPE 50 HEAT GRENADE

Chinese Communist Type 50 HEAT Grenade

The Chinese Communist Type 50 HEAT grenade, which is muzzle loaded, is designed to be fired from the Type 56 launcher, a copy of the Soviet RPG–2 recoilless antitank grenade launcher. The complete round has an 80-mm shaped-charge warhead, a base detonating fuze, a 40-mm stabilizer shaft with a fin assembly of six flexible fins, and a black powder propellent charge in a moistureproof cardboard and aluminum container.

Six rounds, with unassembled propellent charges and fuzes, are packed in a wooden box. A canvas carrier for three rounds and accessories is available for a gunner to carry on his back.

CHARACTERISTICS

Using weapon	Type 56 grenade launcher (copy of Soviet RPG–2)
Type	HEAT
Weight	4 lb
Length	25.6 in
Max diameter	80 mm
Color	Brownish-green
No. of fins	6 flexible
Armor penetration	7 in at 0°

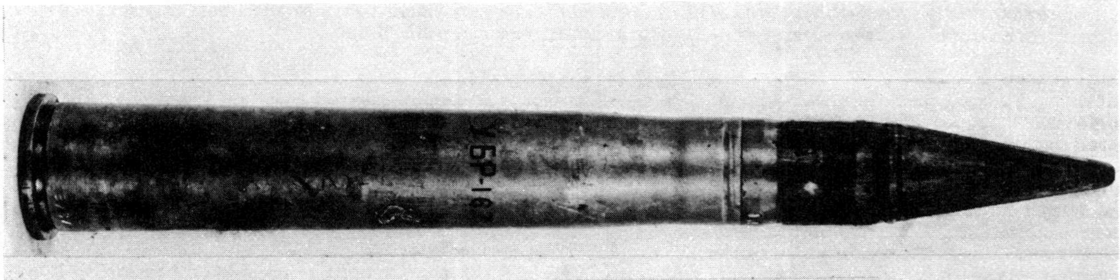

SOVIET 37-mm MODEL UBR-167 AP-T CARTRIDGE

Soviet 37-mm Model UBR–167 AP–T Cartridge

The Soviet 37-mm Model UBR–167 AP–T cartridge has a solid steel projectile with a windshield and with a tracer threaded to its base.

Model UBR–167 ammunition is packed either 30 or 5 rounds in a wooden box. The gross weight of 30 rounds in a wooden box is about 151 pounds.

CHARACTERISTICS

Using weapon	M1939 antiaircraft gun
Caliber	37 mm
Weight	3.3 lb
Overall length	15.1 in
Projectile weight	1.7 lb
Color	Black or unpainted
Case length	9.9 in
Armor penetration	1.8 in at 0° at 500 m

SOVIET 37-mm MODEL UOR-167 FRAG-T CARTRIDGE

Soviet 37-mm Model UOR–167 Frag–T Cartridge

The Soviet 37-mm Model UOR–167 frag–T cartridge is basically designed for use against aircraft but may also be used against personnel. The complete round consists of point detonating fuze with self-destroying element, a steel projectile with high-explosive filler, a tracer assembly, a fixed brass cartridge case, propellant, and primer.

Model UOR–167 ammunition is packed 30 rounds in a wooden box with a gross weight of about 147 pounds.

CHARACTERISTICS

Using weapon	M1939 antiaircraft gun
Caliber	37 mm
Weight	3.2 lb
Overall length	15.1 in
Projectile weight	1.6 lb
Color	Black
Case length	9.9 in

SOVIET 57-mm MODEL UBR-271 AP-T CARTRIDGE

Soviet 57-mm Model UBR–271 AP–T Cartridge

The Soviet 57-mm Model UBR–271 AP–T cartridge has a brass cartridge case (containing single-base propellant and a KV–4 percussion primer) and an armor-piercing projectile with a small explosive charge. The nose of the projectile is covered by a light-gage-steel windshield. The normal fuze for the round is the MD–7 base detonating fuze but others have been substituted.

Five Model UBR–271 rounds are packed in a wooden box with a gross weight of 105 pounds.

CHARACTERISTICS

Using weapons	M1941 and M1943 antitank guns
Caliber	57 mm
Overall length	15 in
Projectile weight	6.9 lb
Color	Olive drab, gray, or unpainted
Case length	18.9 in
Armor penetration	4.17 in at 0° at 500 m

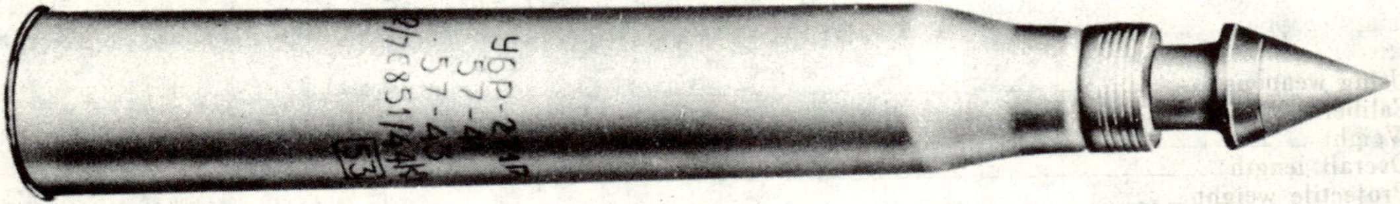

SOVIET 57-mm MODEL UBR-271P HVAP-T CARTRIDGE

Soviet 57-mm Model UBR–271P HVAP–T Cartridge

The Soviet 57-mm Model UBR–271P HVAP–T cartridge has a brass cartridge case and an HVAP arrow-head-type projectile. The cartridge case contains a single-base propellant and is assembled with the KV–4 percussion primer and a loose (unbagged) propellent charge. The projectile has a tungsten carbide core, a tracer element in its base, and an aluminum ballistic cap. Instead of the usual type of copper rotating band, the projectile has a series of mild steel bands and grooves around the cylindrical portion of the body below the waist.

The Model UBR–271P round is packed five to a wooden box, with a gross weight of about 95 pounds.

CHARACTERISTICS

Using weapons	M1941 and M1943 antitank guns
Caliber	57 mm
Weight	12 lb
Overall length	24 in
Projectile weight	3.9 lb
Color	Gray or unpainted
Case length	18.9 in
Armor penetration	5.51 in at 0° at 500 m

SOVIET 57-mm MODEL UO-271U FRAG CARTRIDGE

Soviet 57-mm Model UO–271U Frag Cartridge

The Soviet 57-mm Model UO–271U frag cartridge has an HE-loaded steel projectile, a KTM–1 point detonating fuze, and a brass cartridge case. The fragmentation projectile has a short conical ogive and two copper rotating bands and is held in the cartridge case with a single 360° crimp.

Five unfuzed rounds are packed in a wooden box with a gross weight of about 110 pounds.

CHARACTERISTICS

Using weapons	M1941 and M1943 antitank guns
Caliber	57 mm
Weight	14.5 lb
Overall length	26.5 in
Projectile weight	8.1 lb
Color	Olive drab or unpainted
Case length	18.9 in

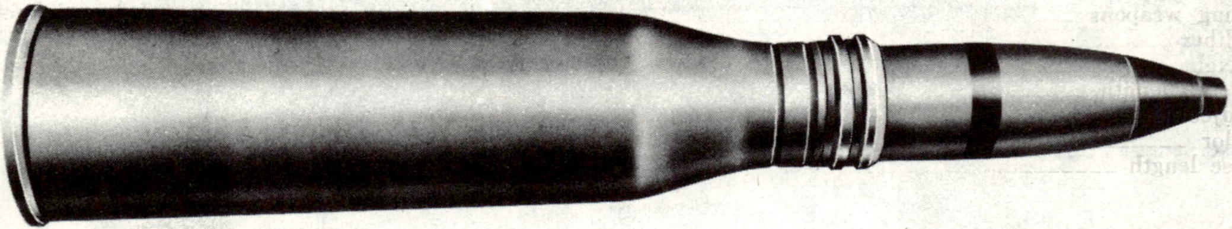

SOVIET 57-mm MODEL UOR-281 FRAG-T CARTRIDGE

Soviet 57-mm Model UOR–281 Frag–T Cartridge

The Soviet 57-mm Model UOR–281 frag–T cartridge is a high-velocity round designed for use against aircraft flying at low altitudes. It can also be employed against personnel or thin-skinned vehicles. The complete round has a point detonating, self-destroying fuze, high-explosive-filled steel projectile, tracer assembly, brass cartridge case, propellant, and percussion primer.

Five Model UOR-281 rounds are packed in a wooden box with a gross weight of about 108 pounds.

CHARACTERISTICS

Using weapons	Models S–60 and S–68 antiaircraft guns
Caliber	57 mm
Weight	14.3 lb
Overall length	21.1 in
Projectile weight	6.3 lb
Color	Olive drab
Case length	13.7 in

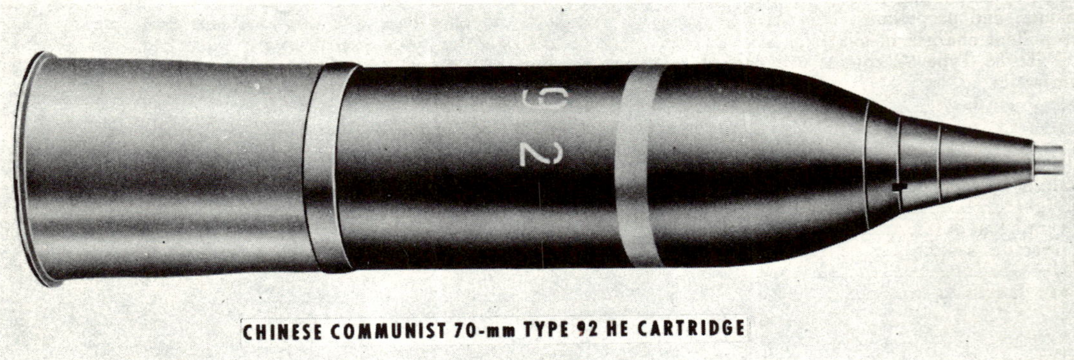

CHINESE COMMUNIST 70-mm TYPE 92 HE CARTRIDGE

Chinese Communist 70-mm Type 92 HE Cartridge

The Chinese Communist 70-mm Type 92 HE cartridge, a copy of a Japanese-designed cartridge, has a Type 88 point detonating fuze, a high-explosive-filled steel projectile, a fuze adapter, a short brass cartridge case with primer, and propellant. The cartridge case fits over the projectile base like that of a semifixed round. Four propellent charges in silk bags and an igniter are placed over the percussion primer.

Three Type 92 rounds are packed in a wooden box with fuzes packed in separate containers near the projectile.

CHARACTERISTICS

Using weapon	Type 92 howitzer
Caliber	70 mm
Weight	9.3 lb
Overall length	14.5 in
Projectile weight	8.1 lb
Color	Blue-green
Case length	3.99 in

GERMAN (WORLD WAR II) 75-mm AP-T CARTRIDGE

German (World War II) 75-mm AP–T Cartridge

The German (World War II-design) 75-mm AP–T cartridge has an armor-piercing projectile with high-explosive filler, a base detonating fuze, and a primed brass cartridge case containing double-base propellant. The projectile has a capped armor-piercing core and a windshield welded to the cap. The base detonating fuze, which incorporates the tracer, detonates the small amount of RDX in the projectile.

CHARACTERISTICS

Using weapon	M40 (Pak 40) antitank gun
Caliber	75 mm
Weight	26.8 lb
Overall length	38.1 in
Projectile weight	15 lb
Color	Black
Case length	28.1 in
Armor penetration	3.23 in at 30° at 1000 m

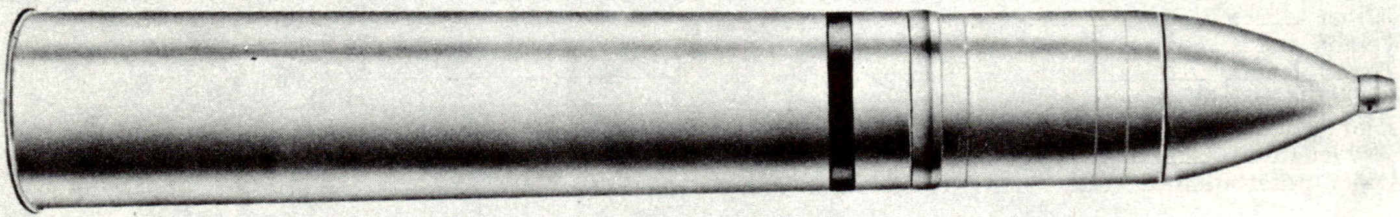

SOVIET 76-mm MODEL UBP-354M HEAT CARTRIDGE

Soviet 76-mm Model UBP–354M HEAT Cartridge

The Soviet 76-mm Model UBR–354M HEAT cartridge has a BM point detonating fuze and a steel projectile containing a steel cone, a paper flash tube, booster, bursting charge, and tracer. The projectile is fitted to a primed brass cartridge case containing the propellent charge.

Model UBP–354M ammunition is packed five, six, or eight rounds in a wooden box. The gross weight of the five-round box is about 128 pounds.

CHARACTERISTICS

Using weapons	M1939 and M1942 field guns; D–56T (PT–76 tank) gun
Caliber	76 mm
Weight	12.6 lb
Overall length	23.7 in
Projectile weight	8.7 lb
Color	Gray or unpainted
Case length	15.1 in
Armor penetration	4 to 5 in at 0°

SOVIET 76-mm MODEL UBR-354B AP-T CARTRIDGE

Soviet 76-mm Model UBR–354B AP–T Cartridge

The Soviet 76-mm Model UBR–354B AP–T cartridge has a base-fuzed AP projectile crimped into a primed brass cartridge case. The projectile is loaded with an high-explosive bursting charge and incorporates an MD–7 detonating fuze, with tracer. The cartridge case contains a multiperforated grain propellent charge loosely loaded into the case and a KV–4 primer. The blunt nose of the projectile is covered by a pointed steel windshield which is crimped into two deep annular grooves machined in the projectile's body forward of the upper bourrelet.

Model UBR–354B ammunition is packed five, six, or eight rounds in a wooden box. The gross weight of the five-round box is about 128 pounds.

CHARACTERISTICS

Using weapons	M1939 and M1942 field guns; D–56T (PT–76 tank) gun
Caliber	76 mm
Weight	20 lb
Overall length	24 in
Projectile weight	14.3 lb
Color	Gray
Case length	15.1 in
Armor penetration	2.72 in at 0° at 500 m

SOVIET 76-mm MODEL UBR-354P HVAP-T CARTRIDGE

Soviet 76-mm Model UBR–354P HVAP–T Cartridge

The Soviet 76-mm Model UBR–354P HVAP–T cartridge has a hypervelocity armor-piercing tracer arrowhead-type projectile held in a brass cartridge case by a 360° ring crimp. The projectile consists of a steel body with a single rotating band, a tungsten-carbide core, a tracer assembly in the base, and an aluminum ballistic cap. The cartridge case contains a short percussion primer and a loosely packed propellent charge.

The Model UBR–354P ammunition is packed five, six, or eight rounds in a wooden box. The gross weight of the five-round box is about 99 pounds.

CHARACTERISTICS

Using weapons	M 1939 and M1942 field guns; D–56T (PT–76 tank) gun
Caliber	76 mm
Weight	13 lb
Overall length	21 in
Projectile weight	6.7 lb
Color	Olive drab or gray
Case length	15.3 in
Armor penetration	3.62 in at 0° at 500 m

SOVIET 76-mm MODEL UO-354AM FRAG CARTRIDGE

Soviet 76-mm Model UO–354AM Frag Cartridge

The Soviet 76-mm Model UO–354AM frag cartridge has an HE-loaded projectile force with KTM–1 fuze fitted to a primed brass cartridge case. The projectile has a single rotating band and two bourrelets. The cartridge case contains loose propellant, a KV-4 primer, and a perforated paper flash tube.

The Model UO–354AM ammunition is packed five, six, or eight rounds per wooden box, with gross weights of about 128, 154, and 207 pounds, respectively.

CHARACTERISTICS

Using weapons	M1939 and M1943 field guns; D–56T (PT–76 tank) gun
Caliber	76 mm
Weight	19.8 lb
Overall length	25.3 in
Projectile weight	13.6 lb
Case length	15.1 in

SOVIET 85-mm MODEL UBR-365 AP-T CARTRIDGE

Soviet 85-mm Model UBR–365 AP–T Cartridge

The Soviet 85-mm Model UBR–365 AP–T cartridge has a primed brass cartridge case containing an uncapped armor-piercing projectile. The projectile is fitted with a base detonating fuze and a tracer element. The propellant is packed loosely in a bag, at the base of which is a black powder igniter charge contained in a separate plastic pad.

Model UBR–365 ammunition is packed either three or four in a wooden box. The gross weight of the three-round box is about 139 pounds.

CHARACTERISTICS

Using weapons	KS–12 (M1939) and KS–18 antiaircraft guns, D–44 and SD–44 field guns, and ZIS S–53 (M1944) tank gun
Caliber	85 mm
Weight	35 lb
Overall length	32.5 in
Projectile weight	20 lb
Color	Gray or unpainted
Case length	24.8 in
Armor penetration	4 in at 0° at 1000 m

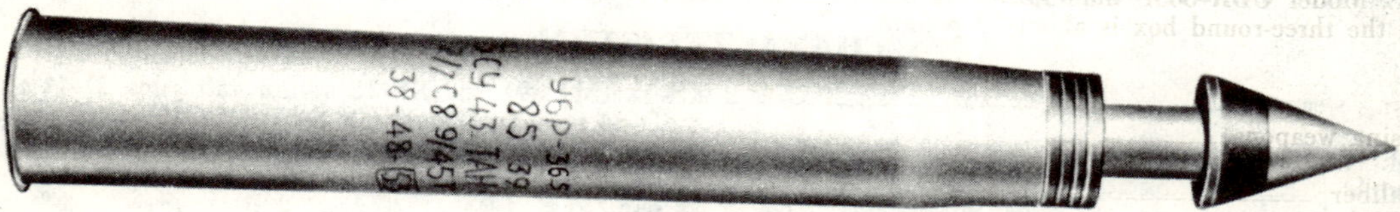

SOVIET 85-mm MODEL UBR-365P HVAP-T CARTRIDGE

Soviet 85-mm Model UBR-365P HVAP–T Cartridge

The Soviet 85-mm Model UBR–365P HVAP–T cartridge has an armor-piercing projectile crimped within a primed brass cartridge case. The steel projectile, which contains a tracer element and a tungsten carbide core, has an arrowhead appearance because of an aluminum nose on its forward portion. The cartridge case contains a percussion primer and a propellent charge packed loosely in a bag. A quilted section on the bottom of the bag contains a loose black powder igniter charge.

Model UBR–365P ammunition is packed either three or four rounds in a wooden box. The gross weight of the three-round box is about 97 pounds.

CHARACTERISTICS

Using weapons	KS–12 (M1939) and KS–18 antiaircraft guns; D–44 and SD–44 field guns; ZIS S–53 (M1944) tank gun
Caliber	85 mm
Weight	26 lb
Overall length	34 in
Projectile weight	11 lb
Color	Gray or unpainted
Case length	25 in
Armor penetration	5.1 in at 0° at 1000 m

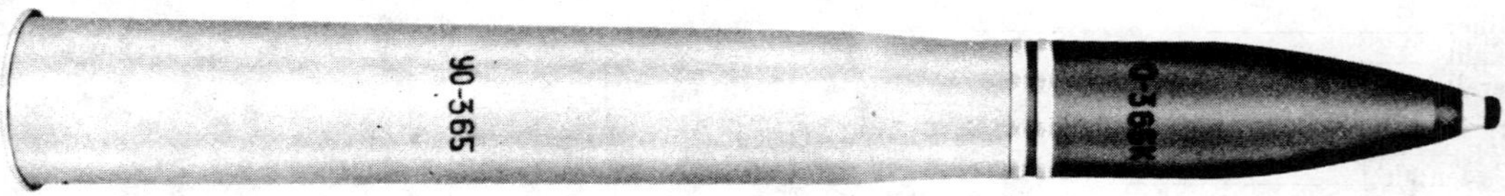

SOVIET 85-mm MODEL UO-365 FRAG CARTRIDGE

Soviet 85-mm Model UO–365 Frag Cartridge

The Soviet 85-mm Model UO–365 frag cartridge has a conventional brass cartridge case and a high-explosive-filled steel projectile fitted with a KTM–1 fuze. The propellant is initiated by a black powder igniter pad and a KV–4 primer. The projectile is held in the cartridge case by a 360° roll crimp by using the lower of two cannelures.

Model UO–365 ammunition is packed three or four rounds in a wooden box. The gross weight of the four-round box is about 167 pounds.

CHARACTERISTICS

Using weapons	KS–12 (M1939) and K–18 antiaircraft guns; D–44 and SD–44 field guns; ZIS S–53 (M1944) tank gun
Caliber	85 mm
Weight	35.4 lb
Overall length	37 in
Projectile weight	21 lb
Color	Gray
Case length	24.8 in

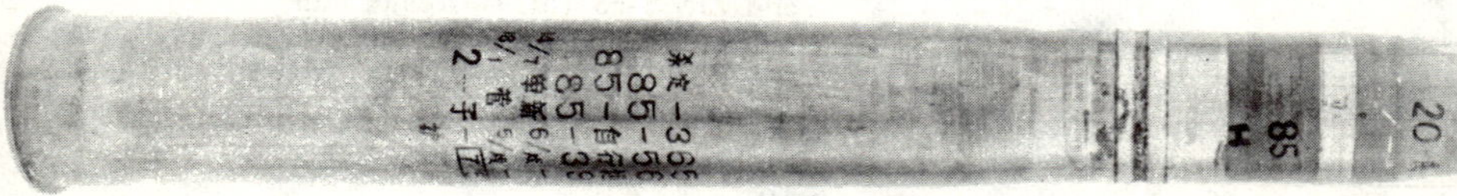

CHINESE COMMUNIST 85-mm TYPE 365 FRAG CARTRIDGE

Chinese Communist 85-mm Type 365 Frag Cartridge

The Chinese Communist 85-mm Type 365 frag cartridge, a copy of the Soviet Model UO–365K frag cartridge, has a primed brass cartridge case and a one-piece high-explosive-filled steel projectile. The projectile is fixed in the case with two 360° crimps. Impact or time fuzes may be used with this round.

CHARACTERISTICS

Using weapons	Type 56 antiaircraft gun; Soviet KS–12 (M1939) and KS–18 antiaircraft guns, D–44 and SD–44 field guns, and ZIS S–53 (M1944) tank gun
Caliber	85 mm
Weight	34.7 lb w/o fuze
Overall length	35.8 lb w/o fuze
Color	Olive drab
Case length	25.8 in

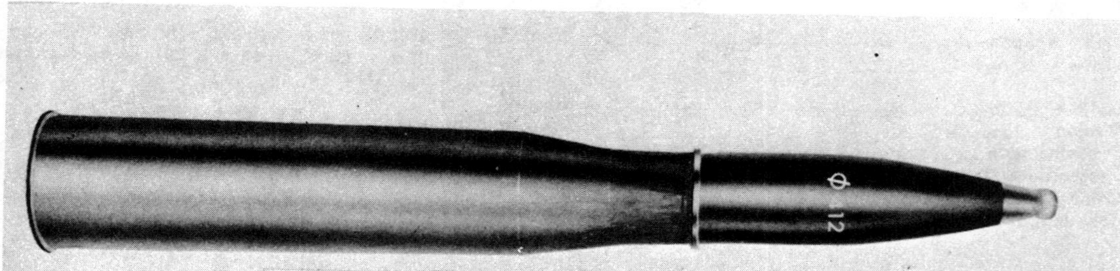

SOVIET 100-mm MODEL UF-412 HE CARTRIDGE

Soviet 100-mm Model UF–412 HE Cartridge

The Soviet 100-mm Model UF–412 HE cartridge has a point detonating fuzed projectile and a brass cartridge case. The projectile is fixed to the cartridge case with two 360° crimps, one at the mouth of the case and the other approximately 1 inch below the first. This round may also be found with a time fuze.

Model UF–412 ammunition is packed two in a wooden box with a gross weight of about 180 pounds.

CHARACTERISTICS

Using weapons	KS–19 antiaircraft gun; Soviet M1944 field gun, D–10T tank gun, and D–10TS assault gun
Caliber	100 mm
Weight	66 lb
Overall length	42.5 in
Projectile weight	35 lb
Color	Olive drab
Case length	27.4 in

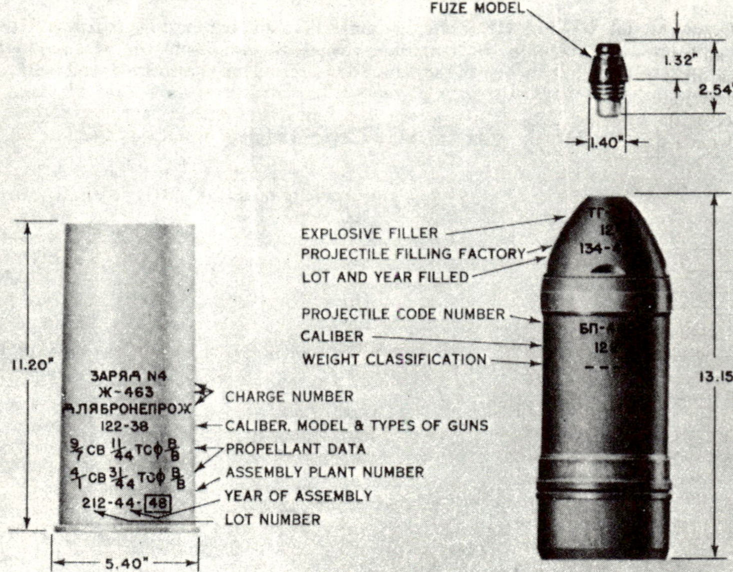

SOVIET 122-mm MODEL VBP-463A HEAT CARTRIDGE

Soviet 122-mm Model VBP-463A HEAT Cartridge

The Soviet 122-mm Model VBP–463A HEAT cartridge is separate loading. It has a two-piece projectile, with a V-229 fuze and a bursting charge with detonator booster assembly in the base, assembled to a cartridge case of either brass or spiral wrapped steel filled with four increment charges of propellant. A KV-4 primer is assembled to the cartridge case.

Model VBP–463A ammunition is packed in a wooden box either with 2 complete rounds or separately with 3 projectiles or 20 cartridge cases. The gross weight of two complete rounds packed in a wooden box is 108 pounds.

CHARACTERISTICS

Using weapon	M1938 (M–30) howitzer
Caliber	122 mm
Weight	37.8 lb
Projectile weight	29.2 lb
Length	14.5 in
Color	Unpainted
Case length	11.2 in
Armor penetration	7 to 8 in at 0°

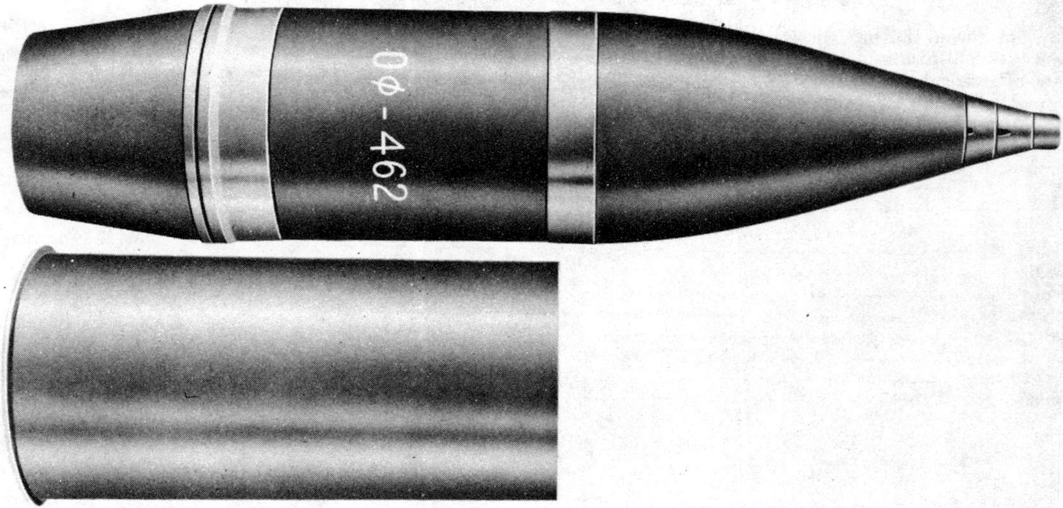

SOVIET 122-mm MODEL VOF-463 FRAG-HE CARTRIDGE

Soviet 122-mm Model VOF–463 Frag–HE Cartridge

The Soviet 122-mm Model VOF–463 frag–HE cartridge is separate loading. It has a point detonating fuzed projectile and a brass cartridge case containing seven bagged propellent charges and an igniter bag. A KV–4 primer is threaded in the base of the cartridge case.

Two Model VOF–463 rounds are packed in a wooden box with a gross weight of 148 pounds.

CHARACTERISTICS

Using weapon	M1938 (M–30) howitzer
Caliber	122 m
Weight	59.8 lb
Projectile weight	47 lb
Color	Olive drab or unpainted
Case length	11.2 in

SOVIET MODEL TMB-2 ANTITANK MINE

Soviet Model TMB–2 Antitank Mine

The Soviet Model TMB–2 antitank mine is constructed from two cardboard pots joined with black tape and coated with asphalt and has a glass stopper on top. Pressure on the mine activates an MV–5 fuze under the glass stopper and explodes the mine. The TMB–2 antitank mine is brownish in color.

CHARACTERISTICS

Type	Blast
Max diameter	10.8 in
Height	5.3 in
Weight	15.4 lb
Actuating force	26 lb
Case material	Cardboard
Case thickness	0.38 in
No. of fuze wells	1 main
No. of serrations	None
Main charge	Amatol or TNT
Filler weight	11 to 14.5 lb
Fuze model	MV–5
Fuze type	Pressure
Safety device	Safety pin

SOVIET MODEL TM-41 ANTITANK MINE

Soviet Model TM–41 Antitank Mine

The Soviet Model TM–41 antitank mine is a cylindrical blast-type mine with an MV–5 pressure fuze inside a sheet steel case. It is easily detected with electronic mine detectors. Corrugations on the upper part of the case's side wall control the way in which the case crushes when a vehicle or tank runs over it and also insure reliable functioning of the pressure fuze.

CHARACTERISTICS

Type	Blast
Max diameter	10 in
Height	5.8 in
Weight	11.9 lb
Actuating force	440 lb
Case material	Sheet steel
Case weight	3.3 lb
No. of fuze wells	1 main
No. of serrations	None
Main charge	TNT or amatol
Filler weight	8.4 lb
Fuze model	MV–5
Fuze type	Pressure

SOVIET MODEL TM-46 ANTITANK MINE

Soviet Model TM–46 Antitank Mine

The Model TM-46 is the latest known Soviet antitank mine. This TMN-46 version has a second secondary fuze well located in the bottom of the case. This mine may be emplaced either by hand or by mechanical means. A pressure of about 400 pounds applied to the pressure plate compresses the striker spring in the fuze until the striker-retaining ball escapes into a recess in the pressure cap, thus releasing the spring-loaded striker.

CHARACTERISTICS

Type	Blast
Max diameter	12.2 in
Height	2.9 in
Weight	19.2 lb
Actuating force	Approx 400 lb
Case material	Sheet steel
Case weight	6.6 lb
No. of fuze wells	2
No. of serrations	None
Main charge	TNT
Filler weight	12.6 lb
Fuze model	MV5 or MV5K
Fuze type	Pressure
Safety device	None

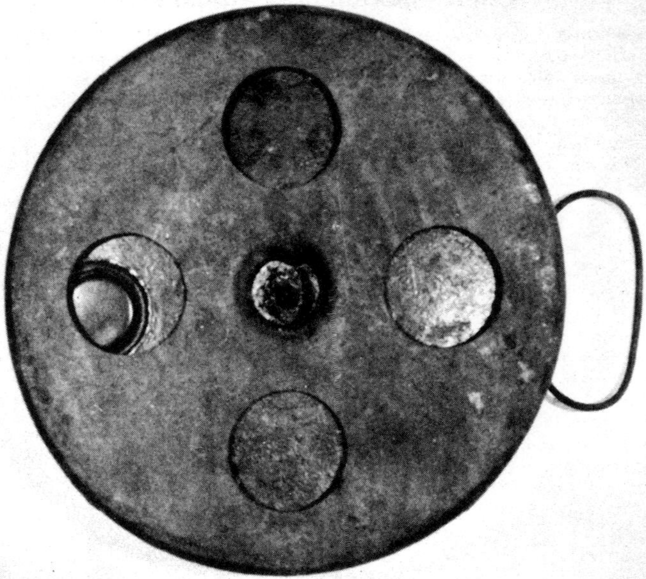

CHINESE COMMUNIST ANTITANK MINE

Chinese Communist Antitank Mine

The Chinese Communist antitank mine is copied from the U.S. M1-series antitank mines. This mine, with a case of very thin sheet steel, has provisions for attaching the pressure plate or spider. The pressure plate, or spider, is a sheet metal disk with four 1.75-inch-diameter holes located 90° apart in the plate surface. Four lugs welded to the outer edge of the pressure plate 90° apart prevent the plate from misalining with the body when the mine is being emplaced. These lugs also act as guides for the downward movement of the pressure plate when a load is applied. All joints are lapwelded, and the inside of the mine case is sprayed with an asphalt-type paint. This flat cylindrical mine, painted olive drab, has no other markings or identifying features.

CHARACTERISTICS

Type	Blast
Max diameter	7.88 in
Height	2.88 in
Weight	11 lb w/pressure plate and fuze
Case material	Sheet steel
Case thickness	0.08 in
Case weight	5.75 lb
No. of fuze wells	1 main
No. of serrations	None
Main charge	TNT
Filler weight	3.5 lb
Fuze type	Pressure
Safety device	Fork

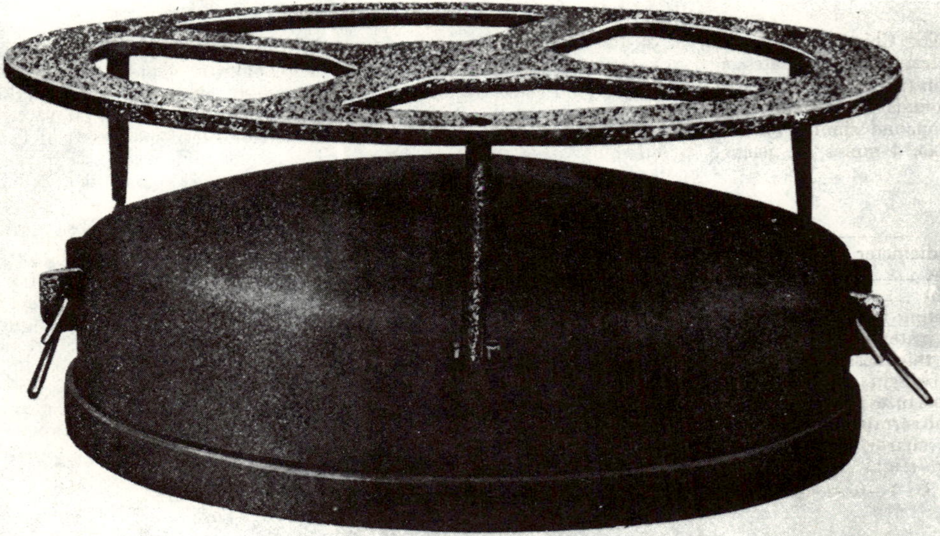

CHINESE COMMUNIST No. 4 DUAL-PURPOSE MINE

Chinese Communist No. 4 Dual-Purpose Mine

The Chinese Communist No. 4 dual-purpose mine was designed for use against both personnel and light vehicles. It incorporates a double-acting fuze which will initiate the explosive charge under either of two circumstances: when a load of 300 to 500 pounds is applied to the pressure spider or when a pull of 10 to 50 pounds is exerted on a tripwire fastened to the fuze's striker-retainer pin. The mine is rustproofed with a compound similar to creosote, and the joint between the two halves of the case is caulked.

No. 4 mines are packed in individual metal carrying cases.

CHARACTERISTICS

Type	Blast
Max diameter	9 in
Height	4 in
Weight	Approx 11.4 lb
Actuating force	300 to 500 lb
Case material	Cast iron
Case thickness	0.1 in
Case weight	5.6 lb
No. of fuze wells	1 main
No. of serrations	None
Main charge	Flake TNT
Filler weight	Approx 4 lb
Fuze type	Pressure or pull

APPENDIX I. CHINESE CHARACTERS

Part 1. Script Form

A number of common Chinese characters are shown in their script form below.

Chinese	*Romanization*	*Translation*
一	i	1
二	erh	2
三	san	3
四	ssu	4
五	wu	5
六	liu	6

Chinese	*Romanization*	*Translation*
七	ch'i	7
八	pa	8
九	chiu	9
十	shih	10
十一	shih i	11
十二	shih erh	12
十三	shih san	13

Chinese	*Romanization*	*Translation*
二十	erh shih	20
二十一	erh shih i	21
二十二	erh shih erh	22
三十	san shih	30
四十	ssu shih	40
一百	i pai	100
一百零二	i pai ling erh	102
一百一十	i pai i shih	110
一百一十二	i pai i shih erh	112

Chinese	*Romanization*	*Translation*
一百二十二	i pai erh shih erh	122
二百	erh pai	200
一千	i ch'ien	1,000
彈藥	Tan yao	Ammunition
兵工廠	Ping-kung-ch'ang	Arsenal
砲兵	P'ao-ping	Artillery

Chinese	*Romanization*	*Translation*
彈底信管	Tan-ti hsin-kuan	Base detonating fuze
短鎗	Tuan-ch'iang	Carbine
鎗彈	Ch'iang-tan	Cartridge
藥筒	Yao-t'ung	Cartridge case
雙用信管	Shuang-yung hsin-kuan	Combination fuze
全彈	Ch'uan-tan	Complete round

Chinese	Romanization	Translation
倣	Fan	Copy
立方公分	Li-fang kung-fen	Cubic centimeter
炸藥	Cha-yao	Explosive
工廠	Kung-ch'ang	Factory
信管	Hsin-kuan	Fuze
總重	Tsung-chung	Gross weight

Chinese	*Romanization*	*Translation*
砲	P'ao	Gun
輕放	Ching-fang	Handle with care
破壞藥	P'o-huai-yao	High explosive
榴彈砲	Liu-tan-p'ao	Howitzer
碰炸信管	P'eng-cha hsin-kuan	Impact fuze
燒夷彈	Shao-i-tan	Incendiary shell

Chinese	*Romanization*	*Translation*
步兵	Pu-ping	Infantry
瞬發信管	Shun-fa hsin-kuan	Instantaneous fuze
防潮	Fang-chao	Keep dry
公斤	Kung-chin	Kilograms
機関鎗	Chi-kuan-ch'iang	Machinegun
月	Yueh	Month

Chinese	Romanization	Translation
迫擊炮彈	P'o-chi-p'ao tan	Mortar shell
山炮	Shan p'ao	Mountain gun
手鎗	Shou-ch'iang	Pistol
彈頭信管	Tan-t'ou hsin-kuan	Point detonating fuze
爆管	Pao-kuan	Primer
彈頭	Tan t'ou	Projectile

Chinese	*Romanization*	*Translation*
激射藥	Chi-she-yao	Propelling charge
步鎗	Pu-ch'iang	Rifle
火箭	Huo-chien	Rocket
炮彈	P'ao-tan	Shell
藥筒	Yao-t'ung	Shell case
短時延期信管	Tuan-shih yen-ch'ih hsin-kuan	Short delay fuze

Chinese	*Romanization*	*Translation*
榴霰彈	Liu-hsien-tan	Shrapnel
煙藥彈	Yen-mu-tan	Smoke shell
定時信管	Ting-shih hsin-kuan	Time fuze
茶褐火藥	Ch'a-ho-huo-yao	TNT
式	Shih	Type
體積	T'i-chi	Volume
年	Nien	Year

Part 2. Stencil Form

Common Chinese characters which either apply or are peculiar to small arms ammunition are listed below in their stencil form.

Ordinary (ball bullet)

Light (type of ball bullet)

Armor piercing

Incendiary

Tracer

Steel (bullet core material)

Brass (cartridge case material)

Iron (cartridge case material)
[material is usually steel]

Type (model)

APPENDIX II. CYRILLIC LETTERS

The following is a list of Cyrillic letters that appear in the prefix of Soviet model numbers for artillery, mortar, and infantry antitank launcher rounds.

Cyrillic	*English*	*Prefixes*
у	U	Fixed complete round
В	V	Separate loading complete round
А	A	Propaganda
Б	B	Armor piercing
БЗ	BZ	Armor piercing incendiary
БЗР	BZR	Armor piercing incendiary tracer
БР	BR	Armor piercing tracer
Г	G	Concrete piercing
Д	D	Smoke
З	Z	Incendiary
О	O	Fragmentation
ОЗ	OZ	Fragmentation incendiary
ОР	OR	Fragmentation tracer
Оф	OF	Fragmentation high-explosive
ОфР	OFR	Fragmentation high-explosive tracer
ОХ	OKh	Fragmentation gas
Пу	PU	Target practice
Р	R	Tracer[1]
С	S	Illuminating (star)
ф	F	High-explosive
Х	Kh	Gas
Бк	BK	High-explosive antitank (HEAT) fin-stabilized ammunition
БП	BP	High-explosive antitank (HEAT) artillery ammunition
ПГ	PG	High-explosive antitank (HEAT) launched grenades and rockets
кГ	KG	High-explosive antitank (HEAT) (hand grenades)

[1]With small arms ammunition, the presence of tracer is indicated by a letter "T" in the marking.